Praise for *This Beautiful Country*

'This marvellous, kaleidoscopic paean to a world of Italian nature that is nearly unknown is divine. While I might have considered Edward's start-up salamander of a speckled sort, the pale, quivering olm, night-black squirrels or violent gladiator crabs to be fantasies, I have visited his world and truly believe. This book is a must-buy for travellers and dreamers or any naturalists who wish to be wary of porcupine quills in the sand. I, for one, will be revisiting soon.'

DEREK GOW, author of *Hunt for the Shadow Wolf*

'From the bubbling quadruple summit craters of Mount Etna to the subterranean weirdness of the sightless, lungless, ancient olm, Edward Cutler takes the reader through a kaleidoscopic compendium of Italian natural history and wonderment. He has sparked in me a fresh and lifelong interest and love for his beautiful home country. This is a truly magical and important book.'

MERLIN HANBURY-TENISON, author of *Our Oaken Bones*

'In *This Beautiful Country*, Edward Cutler describes in epic, beautiful detail the unique natural wonders of his adopted country, Italy.'

BEN GOLDSMITH, author of *God Is an Octopus*

'A deep dive into how hundreds of millions of years of evolution and Earth history have shaped the beauty of modern-day Italy. This engaging and enlightening book will give you a whole new perspective on how one of the world's most fascinating countries and cultures came to be.'

STEVE BRUSATTE,
University of Edinburgh palaeontologist; *New York Times* bestselling author of *The Rise and Fall of the Dinosaurs*

'Probably the best book on the natural and cultural history of Italy since Pliny the Elder.'

CLIVE OPPENHEIMER, author of *Mountains of Fire*

'Edward traces Italy's deep and extraordinary food and agricultural traditions, going back to its geological and evolutionary history, and shows how the land itself shaped what grows and how people eat. I've always known, through the work we do at Natoora, that Italy gives us more diversity, more fruits and vegetables, than anywhere else we source from. It was an instinct, yet *This Beautiful Country* opened that understanding in a way I wasn't prepared for. Reading it gave depth to a feeling I've carried for years, that quiet certainty of why we return to Italy again and again. For anyone interested in food, culture and the importance of biodiversity to the health of our planet, this book is a must.'

FRANCO FUBINI, founder and CEO, Natoora;
author of *In Search of the Perfect Peach*

'If you want to really know Italy, this beautiful, enchanting, witty book is as essential as a visit to the Uffizi or a plate of cacio e pepe in Rome. It reveals the real *bella* Italia for the first time.'

TIM FLANNERY, author of *Europe: The First 100 Million Years*

'With a rigorously documented perspective, *This Beautiful Country* guides readers through the immense richness of Italy's biological and cultural diversity – species, ecosystems, landscapes, arts, customs, dialects, culinary traditions and more. It is a love letter to the country Cutler has chosen as his home, an erudite exploration across multiple disciplines, a unique field guide, and a heartfelt reflection on past human history and the present challenges of preserving this extraordinary heritage. It is also a gentle but effective call to action. Essential reading for anyone who appreciates the privilege of living in Italy.'

LUIGI BOITANI, professor emeritus of conservation biology,
Sapienza University of Rome

This Beautiful Country

This Beautiful Country

How *Italy's Nature* Shaped Its Culture and Civilisation

EDWARD CUTLER

Chelsea Green Publishing
White River Junction, Vermont
London, UK

First published in 2026 by Chelsea Green Publishing | PO Box 4529 | White River Junction, VT 05001 | West Wing, Somerset House, Strand | London, WC2R 1LA, UK | www.chelseagreen.com
A Division of Rizzoli International Publications, Inc. | 49 West 27th Street | New York, NY 10001 | www.rizzoliusa.com

Publisher: Charles Miers
Deputy Publisher: Matthew Derr
Commissioning Editor: Muna Reyal
Project Manager: Susan Pegg
Copy Editor: Susan Pegg
Proofreader: Jacqui Lewis
Indexer: Charmian Parkin
Designer: Jenna Richardson

ISBN 978-1-91529-476-0 (hardcover) | ISBN 978-1-91529-477-7 (ebook) | ISBN 978-1-915294-78-4 (audiobook)
Library of Congress Control Number: 2026018440 (print)
A CIP catalogue record for this book is available from the British Library.

Our Commitment to Green Publishing
Chelsea Green sees publishing as a tool for cultural change and ecological stewardship. We strive to align our book manufacturing practices with our editorial mission and to reduce the impact of our business enterprise in the environment. We print our books using vegetable-based inks whenever possible. This book may cost slightly more because it was printed on paper from responsibly managed forests, and we hope you'll agree that it's worth it. *This Beautiful Country* was printed on paper supplied by TJ Books that is certified by the Forest Stewardship Council.

Authorized EU representative for product safety and compliance
Mondadori Libri S.p.A. | www.mondadori.it
via Gian Battista Vico 42 | Milan, Italy 20123

Printed and bound in Great Britain by TJ Books, Padstow, Cornwall.
10 9 8 7 6 5 4 3 2 1 26 27 28 29 30

MIX
Paper | Supporting responsible forestry
FSC® C013056

For my parents, Tim and Christine

There should be a history of this Italy, a history outside the framework of time, confiding itself to that which is changeless and eternal, in other words, a mythology. This Italy has gone its way in darkness and silence, like the earth, in a sequence of recurrent seasons and recurrent misadventures. Every outside influence has broken over it like a wave, without leaving a trace.

Carlo Levi, *Christ Stopped at Eboli*

Contents

INTRODUCTION

Salamandrina

> What induces you, O man, to depart from your home in town, to leave parents and friends, and go to the countryside over mountains and valleys, if it is not the beauty of the world of nature?
>
> **Leonardo da Vinci**, *Notebooks*

'Papa, is there something in the water over there, near that rock?'

'No. Probably just a leaf or something.'

'I saw it move... Looks kind of like a lizardy thing.'

I squinted at the still water of the Merloncione, a shallow seasonal stream that runs along the bottom of a steep, forested gorge separating our hill from the village of Pereta in southern Tuscany. Mostly hidden by foliage, the stream is more often heard than seen. After heavy rain, you can sit on the village walls and listen to the torrent far below; a low, rhythmic pulse echoing through the rock cleft, the water feeling its way around the contours of the hill on which this remarkable little village is perched. For much of the year, the Merloncione barely rises above a trickle. But time and neglect have dammed it here and there with fallen boulders and unchecked vegetation, creating a string of small, persistent pools. These hold whatever water the valley gathers, forming quiet refuges that remain through the arid heat of summer.

It was one of the first places we'd explored after settling in our new home. On the village side, we'd discovered the rusting carcass of a Fiat 500, an abandoned relic from another time, most likely dumped from the road above one night by someone not wishing to pay the scrappage charge. As a child, I had often fantasised about pushing a car off a cliff, and I sometimes wondered which of the elderly villagers, perched on their benches, might be hiding this naughty little secret.* Down below in its final resting place, this 'car of the people', a symbol of Italy's post-war industrial boom, had slowly merged with the wild riparian environment, its shape nearly unrecognisable, wrapped in vines and creeping blue periwinkle. On a good day, I might have seen it as an Arcadian touch – a trace of civilisation peacefully returning to nature. On more sombre days, it was just another reminder of humanity's habit of treating the planet like a dumping ground.

Apart from a rusting car, the odd deer, a scattering of deadly mushrooms or the occasional intriguing track and pile of poo, we rarely found much of interest in our stream. More than once, feeling virtuous, we filled black sacks with the trash washed down by gravity and carelessness from above, caught and decaying among the rocks and roots that laced the water's edge. Yet it hardly mattered. The sound of running water and the stillness of the surrounding woodland were enough to make the place one of adventure and

---

* For the record, I have never pushed a car off a cliff.

significance for us. But on that early April day in 2020, the Merloncione became something more than a familiar spot. In a shallow, calm reach of the stream, I leaned closer to the water where my son was pointing. At once, I saw it – an amphibian-shaped shadow, hovering delicately in the current, poised in the shadow of a rock, completely unperturbed by our presence above. Feeling a genuine rush of discovery, I congratulated my son on his sharp eye. We both knew we had found something unusual. Clearly, the moment deserved to be captured with a proper identification, so I looked around for help.

Having found a discarded jam jar and recalling the example of a biology teacher and her lab aquarium at school years before, I gently submerged our receptacle with the opening in front of the shadowy creature. Tipping it down and then jerking the jar up, I gently wafted the creature inside. It was almost too easy – so much so I feared it might be dead. Carefully lifting it back out of the water, we saw that it was very much alive, displaying a speckled black-and-white belly, with the vivid red undersides of its legs and tail splayed against the glass. Looking closer, I saw it also had distinctive flashes of white at its mouth and two tiny white dashes above the eyes on an otherwise completely black topside, almost the photo negative of someone wearing a little Halloween skeleton costume.

We were elated with our discovery and, like any father who tries to sound knowledgeable to his young son, I did my best to describe a creature I freely admitted I had never seen before, but which would most likely be some sort of newt or other amphibian. After obtaining photographic evidence and returning it to its original spot under the rock, a simple search later introduced us to the remarkable, and aptly named, spectacled salamander, a creature that would come to dominate my thoughts over the following years.

Beyond all the biological information regarding its life cycle, what stood out was its endemic status to the Apennines, the only place in the world it could be found. I felt thrilled and a deep sense of local pride at the idea of hosting a unique amphibian in our area. I was further delighted to learn that, like human Italians, the spectacled salamander is distinguished by being either a northerner like ours, *Salamandrina perspicillata*, or a southerner, *Salamandrina terdigitata*, who are almost impossible to tell apart by sight but divided by genetic variations as well as the physical obstacle of the Volturno river. For humans, the north–south divide in Italy is usually marked by the latitudinal waypoints of either the Spezia–Rimini or Massa–Senigallia lines, depending on who you ask. I myself prefer to draw it at the Brioche–Cornetto Line, depending on what you call your breakfast pastry.

I was especially proud to have shared that moment with my son, who, at that age, was still brimming with enthusiasm and curiosity for the world around him. It brought back flashes of my own childhood in rural Suffolk: pond-dipping, falling into streams, roaming the woods with the kind of wide-eyed wonder that seems to bloom only once in a lifetime. Much of that countryside I had known so well has since been stripped back and tidied into the featureless arable landscapes that now dominate much of Britain – hedgerows flailed to stumps, margins drained, the soft disorder that once sheltered life shaved away for efficiency. It remains one of the great paradoxes of the UK: a nation of self-professed nature lovers presiding over one of the most depleted ecologies in Europe.

Yet it is precisely from this point of loss that the country has begun to take bold action. Over the past two decades, a wave of rewilding projects and nature-friendly farms has shown how quickly damaged landscapes can recover when simply given the chance. From Knepp to Wild Ken Hill, Alladale in Scotland to Cabilla in Cornwall, the concerted efforts of rewilders are driving the fastest regeneration of British nature in generations.

Italy seems to suffer from the opposite problem. While in the UK we have had to act urgently in recent years to restore nature, through rewilding and implementing natural capital schemes to subsidise the healing of our depleted lands, Italy has been casually rewilding for years. Forest cover now exceeds 40 per cent of the country, large carnivore numbers, particularly wolves, have returned to levels not seen for centuries, and some of the last true wilderness areas in Europe are being conserved. Much of this can be attributed to demographic shifts, with over 70 per cent of the population now living in urban centres, particularly the sprawling conurbations of Rome, Milan, Naples and Turin. The remaining 17 million people are spread across Italy's varied hilly and mountainous terrain, which 75 per cent of the country is classified as. They live in landscapes that can be harsh and unyielding, often in small towns and villages that barely hold on to their dwindling populations or precariously cling to eroding hilltops. Italy is home to about 6,000 ghost towns and over 15,000 towns where 90 per cent of the population has departed. Abandoned farmland quickly turns to scrub, then eventually back to forest, creating the perfect environment for inland wildlife to flourish. When demographic shifts and rural economic decline meet complex topography, wilderness and nature quickly reclaim the land.

After that first encounter with the *Salamandrina*, my curiosity about Italy's endemic creatures grew. A few simple searches led me to staggering facts and figures about the country's natural heritage, completely upending my understanding of it. I had always known, on some level, that Italy was a wild place, certainly far wilder than anything I had experienced growing up in England. But now, armed with incontrovertible data from government and EU sources as well as curious enthusiast websites, I began to see the full picture: Italy has the highest levels of biodiversity in Europe, with over 60,000 recorded faunal species and an exceptionally high rate of endemism. A significant proportion of its land and freshwater animals are found nowhere else on Earth. The flora tells a similar story: Italy supports around 8,000 native vascular plant species – those with true roots, stems and leaves – and nearly a fifth of them are endemic, among the highest proportions on the continent. Set against continental figures, the scale becomes clearer: Europe counts roughly 150,000–200,000 faunal species and about 15,000 vascular plants, which means Italy alone holds around a third of Europe's animals and half of its plant species.

Italy's Mediterranean neighbours – Greece, Spain and much of the Balkans – are also vitally important repositories of European biodiversity, with similarly high levels of endemism. But I had never imagined that Italy, with its centuries of civilisation building, deforestation, draining, industrialisation, agriculture and urban sprawl, could still be such a vital refuge for the continent's flora and fauna, including some of its most unique species. It was a shock to me, an Italian resident for more than fifteen years, with what I thought was a solid general knowledge of the country's history and cultural heritage. Yet, somehow, I was unaware of this extraordinary reality of Italy's wildlife. It felt like discovering that your best friend, someone you'd known for years, holds a Guinness World Record for a talent they had never revealed. Shouldn't this be common knowledge?

I asked my wife and then my Italian friends whether they were aware of these facts. They were all oblivious and pleasantly surprised when I told them that Italy is, in effect, the Madagascar of Europe, a land apart from its continent in which the flora and fauna have evolved independently; a slight exaggeration, but one that serves well for dramatic effect. It was a relief to know I wasn't alone in my ignorance but also sad to realise that a remarkable creature like the spectacled salamander was probably unfamiliar to most Italians (though admittedly based on a very limited local survey). And it wasn't just the *Salamandrina*; Italy's roll call of endemic life reads like a catalogue of evolution's survivors. The Nebrodi fir, the world's rarest conifer, still grips a single Sicilian ridge, and the Apennine edelweiss blooms only high up on Italy's limestone ledges. The owl moth of Monte Vulture – a remarkable relic from before the Ice Age – flutters its patterned wings alone above volcanic forests. In the clear springs of Lake Posta Fibreno swims a trout found nowhere else on Earth, while the blue lizard of Capri suns itself on sea cliffs no other creature can reach. The ghostly cave salamanders of Sardinia and the underground spaces of the Apennines, breathing through their skin, linger in darkness, while the Nuragic spider spins its silk in the Sardinian scrub. The yellow-bellied toad flashes its warning colours in Apennine puddles after rain; the Calabrian black squirrel and the Marsican brown bear move through beech and oak; and the Apennine chamois picks its way across mountain crags once carved by glaciers.[1]

Of all the remarkable plants and creatures I had suddenly learned about, most seemed to be known primarily through the efforts of dedicated academics and their published papers, along with a handful of amateur naturalist websites and the social media pages of nature photographers. Apart from the more enigmatic bears and wolves, these species were largely absent from the popular books, films or TV documentaries about Italy that I could find. It confirmed what I had already suspected – and what the great biologist Luigi Boitani later put in writing to me: 'Few look at Italy from the point of view of biodiversity: landscapes are celebrated, forgetting that they are made up of ... animal and plant species.'

I arrived in Italy as an art historian by training, with a fledgling career in the art world. Like many others, my knowledge of the country was primarily shaped by a two-and-a-half-thousand-year timeline, starting with the Etruscans and Romans, passing through the Renaissance, the Reformation and the Risorgimento – the unification of Italy as a new country in modern times. We, myself especially, had been looking at Italy's landscape and seeing only its relics and the impressions left by a well-documented dreamscape – a blend of classical history, Renaissance art and architecture, politics, religion, food and wine, fashion, fast cars and football. Aside from those fortunate enough to study Earth sciences here, it seemed that the rest of us had neglected to develop even a basic understanding of Italy's natural history – a country that arguably gave birth to the very subject. Everyone knows about the eruption of Mount Vesuvius that buried Pompeii in 79 CE, but very few understand why there are volcanoes there in the first place.

My career in Italy had taken me from the enclosed art world of Milan to being a roving salesman of green technology and eventually becoming a Tuscan olive farmer. What I was beginning to learn about Italy's remarkable biodiversity, I also began to recognise in its staggering agrodiversity – nowhere else on Earth comes close to cultivating the number of different fruit, vegetable and crop varieties as Italy does. All in a country that covers just 0.5 per cent of the planet's surface.

I decided to go millions of years back in time to try to understand why Italy is the way it is: how the peninsula was formed, what it is made from and how the land has shaped all the life: both the wild and human kind that came to populate it. In Italian political journalism, there's a phenomenon called *dietrologia*, an obsessive analysis of events to uncover the ulterior motives behind the story. Often bordering on conspiracy theory, it's a staple of Italian social commentary, still fuelling the national conversation.

I started to think of my work as a sort of dietrologia – an attempt to look beneath the Italy we think we know, beyond the vineyards and ruins and Renaissance facades, to the forces that made this boot-shaped shard of land a crossroads of peoples, ideas, art and agriculture. And as I followed the country's deep-time story, I began to understand its significance. Italy itself seemed to change shape before my eyes.

The remarkable biodiversity of Italy today is the result of millions of years of isolated evolutionary history at the heart of the Mediterranean and owes much of its existence to the country's unique landscape – one of the most youthful, dynamic and extraordinarily diverse places on Earth. Italy is a land of towering mountains, desert islands, ancient forests, deep cave systems, glaciers, rivers, lakes and Europe's only mainland volcanoes, which emerged from the water barely yesterday, geologically speaking. With such a rich natural playground, it's no coincidence that the study of Earth sciences and natural history, after its initial spark in ancient Greece, first began to properly develop as a science in Italy itself. It felt as though the blinkers had been removed, suddenly seeing the history of Italy unfold over tens of millions of years instead of the few paltry centuries we had studied at school and university. Moreover, this deep-time story I was uncovering seemed to explain and enrich so much about contemporary Italy, offering context for the country's anthropological peculiarities that I had always embraced, if not fully understood.

Just as Italy's striking geodiversity – its mountains, volcanoes, coasts, plains and islands – nurtured one of Europe's richest ecosystems and its biodiversity, it also shaped what we might call anthrodiversity. The peninsula's fractured topography fostered isolation, encouraging distinct human groups to grow apart, each developing its own languages, customs and material cultures. Long before Rome unified the peninsula, Italy was already a mosaic of cultures shaped as much by mountains and valleys as by time. Even today, in some communities of Calabria, one can still hear a form of ancient Greek – a linguistic echo Homer or Plato would have recognised, though modern Greeks would struggle to comprehend. The regional diversity that still characterises Italy has roots deep in this ancient, landscape-driven separation.

My many questions began to be answered over the next couple of years as I immersed myself in research about Italy's endemic wildlife. My investigation widened to cover the country's geology, palaeontology and palaeoclimatology – the study of past climates and how they shaped the present landscape. I came to understand the story of Italy's creation: its complex birth, shaped over the last thirty million years by the slow collision between Africa and Eurasia – forces that continue to mould the country today. These tectonic pressures created a wildly varied landscape – a mountainous amalgam of islands and oceanic crust – that allowed for a vast range of ecological and climatic variation, generating an astonishing number of environmental niches where European fauna found vital refuge during the Ice Age over the last two and a half million years. I also learned how today's biodiversity is just a fragment of the extraordinary creatures that once roamed Italy's islands: pygmy elephants and hippos as recently as fifty thousand years ago on Sicily, and five-horned, vampire-fanged deer, giant dormice and fearsome mammoth hedgehogs on the island of Gargano nine million years ago.

When my son and I were observing our own *Salamandrina*, little did we know that we were gazing into the eyes of a true relic of European fauna. This creature's ancestors evolved around ninety million years ago on an island in a very different Europe, diverging at a time when dinosaurs still roamed the planet – and witnessing their eventual demise. First described in 1821 by zoologist, geologist, museologist and the father of Italian ornithology, Paolo Savi, *Salamandrina perspicillata* (sometimes called Savi's salamander) occupies a particularly symbolic place in the canon of Italy's biodiversity.* The genus *Salamandrina*, established through Savi's work, is today the sole surviving genus of the

* It has been used as the logo for the Italian Zoological Union since 1900.

ancient Salamandrininae subfamily, a living remnant of what palaeontologists like Tim Flannery refer to as Europe's core fauna – those species that have survived from the Mesozoic to the present day.

> *Most of Europe's core fauna is now long extinct, but there are a few unlikely survivors. The most important are the Alytids (the family that includes the midwife toads) and the typical salamanders and newts (family Salamandridae). These relics of Europe's dawning are deserving of special recognition, for they are in effect Europe's living fossils, as precious as the platypus and lungfish.*[2]

Italy has its own endemic relatives of the Alytids. The Apennine yellow-bellied toad, *Bombina pachypus*, is found only south of the Po River at altitude through the Apennines, and the Tyrrhenian painted frog, *Discoglossus sardus*, part of Europe's oldest surviving invertebrate family, is found primarily in Sardinia, Corsica and select islands in the Tuscan Archipelago, including Montecristo.

Tim Flannery's words, from his brilliant book, *Europe: The First 100 Million Years*, calling for a deeper reverence for our ancient relict species, struck a chord with me. How can we truly succeed in our conservation efforts if we don't first cultivate a profound appreciation, deeper understanding and genuine respect for the natural world around us? I had spent years admiring the country's buildings and artworks, which we have collectively eulogised for centuries, but hardly any time reflecting on its natural riches. It became clear to me that we were missing a popular retelling of Italy's natural history. For a country whose inhabitants often disparage its politics, sporting achievements or economic performance, there is, I believe, something worth celebrating in Italy's status as the most biodiverse place in Europe. Above all, Italians should take pride in their country's role as a beacon of European nature, with its conservation efforts arguably ranking alongside all the country's other historical accomplishments.

Over the past 150 years, only two popular natural history books about Italy have been written: Antonio Stoppani's *Il Bel Paese*, published in 1876, not long after the unification of Italy, and Fulco Pratesi's *Storia della Natura d'Italia* from 2001, neither of which had been translated.[3] I wanted to at least try to echo the sentiment of both books, opening our eyes to Italy's natural wealth, but in my own language and showing how deep-time history has shaped the country we see today and how it's still evident all around us. As just one in a very long line of foreigners to have fallen in love and made my home here, it felt like a repayment of sorts, a way of giving back. My aim was to explore and revisit both familiar and unknown places, stripping away the layers of human influence to understand what made them the way they are. I also wanted to appeal to my fellow Italians' sense of pride, just as I had felt when discovering the *Salamandrina* in our local stream.

What follows is not an academic treatise but a work of curiosity and observation, grounded in research, travel and conversation with the many scientists, historians and naturalists whose knowledge underpins it. It is written for anyone who loves, or has loved, Italy for whatever it represents, whether historically, artistically, gastronomically – or simply as it is.

Written by a recovering art historian with an incurable fascination for the ground beneath his feet, this book offers a new telling of Italy's natural history: its geological birth, extraordinary biodiversity and the restless forces that have shaped both land and people. I want to reveal the raw form of the peninsula – the shifting geology that raised its mountains, fed its rivers and determined its soils – and to show how that same landscape gave rise to its cultures, beliefs and ways of life.

The span of this book stretches from the making of the peninsula to the present, though its centre of gravity lies between deep time and the Renaissance – the long era in which Italy's landscapes, species and human imagination first took recognisable form. From there, each chapter rises to the surface of the modern day, tracing how those same forces continue to shape the country we inhabit now. My purpose is not to march through history in sequence, but to uncover the continuities between past and present: the ways in which geology, geography and climate echo through everything from volcanoes and forests to food and faith.

Many of the chapters take single elements as their guide: Stone, Fire, Water, Ice and Wood; others take conditions – Islands, Refuge, Revival and Abundance – each a force that has shaped Italy's landscape and its imagination. Together, they trace how mountains rose, volcanoes burned, rivers cut, forests spread, and people found meaning and shelter within them. Political borders hold little meaning to a geologist or a biologist, and they often blur for a historian too. At times, therefore, this story wanders beyond Italy's frontiers to France, Switzerland, Austria, Slovenia, Croatia and Malta, where the same forces once flowed and the same patterns of life began to take shape.

Naturally, this work is factually incomplete. To wander across more than three hundred million years of history while moving freely between geology, ecology, history and art meant accepting that much would be left out. The chapters are organised thematically rather than chronologically, crossing eras and disciplines in search of connection rather than sequence. Each chapter could easily have grown into a book of its own, and almost as soon as one was finished, another omission appeared. Many subjects resisted confinement to a single element: Ciro, the baby dinosaur of Pietraroja, might just as well have belonged to Islands as to Stone; the Calabrian forest dormouse could have swung across the pages of Wood as much as Ice; the owl moth of Monte Vulture would drift through the realms of Fire, Ice, Wood and Refuge if permitted; while our spectacled salamander, true to form, found a way to slip into almost every chapter. I often found myself in unfamiliar territory, but wherever I went, I met the same generosity and curiosity that animate Italy's scientific and naturalist community. Their kindness kept me afloat whenever I doubted the worth of the task.

Like all countries, Italy wrestles with the relics of its past and the injustices of its present. To write honestly about its landscape is to uncover both beauty and damage, generosity and neglect. I have tried to keep the balance: to praise but not shy away from the reality. Most of the criticisms could apply to many nations; others are uniquely Italian. Ultimately, I remain an Italophile, moved by an enduring wonder for this country's natural and cultural inheritance – so much so that any harm done to its landscape feels all the more grievous.

What began as an account of biodiversity, inspired by a chance encounter with a salamander, has become something far larger: an attempt to understand an adopted homeland beyond its surface. It has been an honour and a privilege to try to write it, although I am fully aware of its shortcomings. I hope readers, especially Italians more knowledgeable than me, will forgive the more glaring errors and omissions. Beneath the postcard layer lies not a dark heart but a molten one: alive, volatile, endlessly creative. Italy is still forming, still surprising us, still teaching us how life and landscape shape one another.

Welcome to This Beautiful Country.

CHAPTER ONE

Stone

And therefore 'Mother' is a name that's fitting
for the Earth.
For what arises from the earth falls back to
earth once more,
And that which was sent down to earth from
heaven's aethereal shore
Is taken up again into the quarters of the sky.

Lucretius, *The Nature of Things*

I decided I needed some perspective, so I did the only thing that made sense and headed for higher ground. The Strada delle Vette – the mountain road leading to the high point of La Sila National Park in Calabria, southern Italy – wound upward through dense pine forests, and though I'd been warned it might be closed or barely passable, the pull of altitude was irresistible. Dinner was still hours away.

The road had some holes and the odd fallen tree branch, nothing too dramatic – it seemed intact, which is not always a given round here. I drove on, eyes fixed ahead for hazards, while in my head I tried to stitch together the story of Italy's rocks. Earlier that day, a helpful guide at the visitor centre in Cupone had valiantly tried to explain the region's bewildering geology to me. I had nodded and made noises suggesting comprehension, but in truth my head was swimming. I was standing at one of the southernmost reaches of the Apennines, and yet technically the rocks beneath my feet had little to do with the rest of Italy's spine, and much more to do with Sardinia and Corsica. 'This is Calabria,' he had told me. 'Everything is different here.'

For non-geologists like me, tracing Italy's story through deep time demands a certain amount of mental gymnastics, but it is worth the effort. Italy's geological diversity, its geodiversity, holds the key to understanding its extraordinary biodiversity, which in turn explains its rich agrodiversity, and ultimately its remarkable anthrodiversity. Everything begins with the rocks.

What little geology I knew came from a few school lessons on Britain and Ireland. On the map it looks almost orderly: neat ribbons of colour sweeping diagonally, each a different age. The youngest bedrock lies in the southeast – chalks and clays in Kent, Sussex and East Anglia, some of it mantled by Ice Age sands and gravels. Inland the rocks grow older: Jurassic limestones in the Cotswolds; Triassic red sandstones in the Midlands; Carboniferous coals in Wales, northern England and central Scotland. Older still, the Silurian and Ordovician slates and volcanics of the Welsh hills and the Lake District, then the Dalradian schists of the Highlands, and at last the Lewisian gneiss of the Outer Hebrides – nearly three billion years old. From those orderly ribbons, one sees why modern geology was born there.

Italy's geological map, by comparison, is psychedelic – a kaleidoscope of colours far less orderly than the neat stripes of Britain. Rocks of wildly different ages lie side by side, stacked and faulted, pierced here and there by red volcanic intrusions, that together tell a story of creation, collision, subsidence, eruption and amalgamation across hundreds of millions of years. The land where I found myself pondering all this, Calabria, perfectly encapsulates the geological madness of Italy. A land of mountains whose roots lie not with the rest of mainland Italy but with Sardinia, and even fragments of the Alps. Perched on the frontline of the ongoing collision between the African and Eurasian tectonic plates, it is seismically restless and geologically alive. Fittingly, this was also the first land to be called Italia.

In La Sila – the vast forested plateau at Calabria's heart – trying to make sense of the rocks requires quite a lot of mental fortitude. Forged deep underground some three hundred million years ago on the southern supercontinent of Gondwana, those rocks were later sheared off as that vast landmass broke apart, welded to Europe, broken again and finally hoisted into their current place when the African tectonic plate bulldozed north. Here, a granite plateau has been grafted onto a limestone chain, a fragment of ancient crust lifted from the depths. To walk across La Sila is to tread the roots of mountains long vanished, survivors of a planetary upheaval that left pieces of Europe stranded beside Africa. Its forests, too, tell of deep time: stands of endemic Calabrian black pine (*Pinus nigra* var. *calabrica*) arise straight from the granite soils, among them the Giganti della Sila, trees over five centuries old and more than 40 metres tall, last remnants of the forest giants that once covered the plateau.

The road eventually straightened and widened, bringing me to the piazza of the Rifugio Botte Donato, a boarded-up 1970s hotel standing glumly below the Lorica cable car arrival station at 1,877 metres up, forming part of a modest ski area. It's easy to forget that in winter one can ski in nineteen of Italy's twenty regions. Beyond the obvious Alpine strongholds of

Piedmont, Lombardy, Trentino-Alto Adige, Veneto and Friuli, skiing extends along the entire Apennine chain and even to the islands of Sardinia and Sicily. The only exception is Puglia, a largely flat limestone plateau, which leaves little room for sustained snow, or meaningful vertical descent.

Nevertheless, a ski resort in the southern Mediterranean still feels incongruous to me, regardless of the altitude, and by mid-June this small Calabrian winter outpost was understandably deserted. Slalom poles and piste markers lay in heaps beside the concrete buildings; the hotel – closed for the season and already sliding back toward dereliction – was shuttered, its upper windows cracked and unhinged. Through the smudged glass of the après-ski bar, I saw the yellow bombardino dispenser sitting forlorn on the counter, a ghost of winter revelry. The entire area felt like a series of clearings carved out of the surrounding forest of Calabrian black pine – trees that seemed ready, given half a chance, to reclaim everything. I walked the last stretch to the top of the runs in total silence, grateful for the cool mountain breeze. Swept by winds from two seas, densely forested and with almost no industry in sight, La Sila is often said to have the cleanest air in Europe. Up here it felt true.

I wandered around the summit, looking up at the great pylon whose wheeled girders carry the gondola cabins into the concrete station. My eyes followed the rungs of the ladder vanishing toward the top. In another life I might have climbed it. Instead, I sat on the ground and gazed out over Lake Arvo, an artificial early twentieth-century hydro reservoir paradoxically poised amid uplands forged more than three hundred million years ago. I thought about how such modern engineering stood among rocks older than time itself, and how I might retell Italy's geological creation in words I myself could understand.

The Old New

In the long history of our 4.5-billion-year-old planet, the land we call Italy appeared only yesterday. It is a youthful apparition, thrust from the sea as the African tectonic plate pressed north and collided with Eurasia. The familiar boot-shaped peninsula has held its outline for scarcely 3 million years, a blink in geological time. As a unified nation it is younger still, little more than a century and a half old, newer even than the US. Yet for all its youth, the human story written upon it is among the most ancient and continuous on Earth, encompassing the Etruscans, Italiotes, Romans and much more. Its landscape, similarly, fuses matter of immense antiquity with startling novelty: rocks formed hundreds of millions of years ago are welded to others born only a few thousand years ago, all crushed together by the restless earth – a land both ancient and new.

Italy's oldest exposed rocks surface in the far southwest of Sardinia, in the Sulcis-Iglesiente region. Pale limestones from the Cambrian period, about 500 million years old, bear witness to the dawn of complex life. Within them lie the fossil remains of trilobites and brachiopods that once thrived in warm, shallow seas thousands of kilometres south of where they rest today. But Sardinia's geology, shared in part with Corsica, is largely distinct from the rest of Italy. Until around 30 million years ago, the two islands were joined to what is now southern France and Spain; they then drifted eastward, slowly pivoting into their present position over 15 million years, helping to push up the Apennines, Calabria and Sicily. This long isolation allowed Sardinia's flora and fauna to evolve along their own paths. Today, among its Palaeozoic rocks, the island shelters hundreds of endemic plant species – about 8 per cent of its native flora – and an array of unique animals: the Sardinian brook newt, the long-eared bat, rare butterflies, beetles, cave salamanders and its own subspecies of red deer, all found nowhere else on Earth.

By contrast, Italy's youngest rocks have formed only within the last few thousand years – and continue to build. The Po Valley, stretching from the western Alps to the Adriatic, is a vast alluvial plain laid down, grain by grain, from sediments still washed down by the rivers that feed it. Known as the Pianura Padana, it is a foredeep basin born of the very mountains that enclose it. Each flood spreads a new skin of silt, sand and clay, quietly raising the land a few millimetres at a time. In geological terms, these newborn rocks are as fresh as wet mortar – evidence that Italy is not the finished article but a land still in the making. Yet even here, among the youngest soils in Europe, evolution has left its signature. Hidden in the remnants of riparian woods and rice-field ditches survives the Italian agile frog (*Rana latastei*), a small, russet amphibian unique to Italy that recalls the primeval floodplains from which this newborn land first emerged.

The Po Valley was among the last parts of Italy to stabilise as land, not through tectonic uplift, but by slow burial beneath its own sediments. Today, it is home to a third of Italy's population and more than half of its agricultural and industrial output. Geologically speaking, it is still subsiding, still in-filling, still being shaped by the mountain ranges that created it. When one looks out across the valley today, with its endless fields, factories and smog, it is hard to imagine that ten thousand years ago this was perhaps the most bio-diverse region in the country: a near-impassable wilderness where the river and its many tributaries wandered through a vast, waterlogged plain, hemmed in by immense riparian forests full of life, but also impenetrable to all but the most determined humans.

Enrosadira

In between these two temporal markers – the prehistoric stones of Sardinia and the ongoing sedimentation of the Po Valley – lies the patchwork of rocks that makes up the rest of Italy, most of them forged over many millions of years and uplifted into place in more recent times. Apart from the Po Valley and the plains of Puglia in the far south, Italy is a land of contours: 70 per cent hills and mountains, squeezed upward by the northward drift of the African tectonic plate as it collided with Eurasia.

For much of this deep time, we must imagine Italy not as a peninsula at all, but as a scattering of coral islands and volcanic atolls adrift in the warm, azure waters of the Tethys. This was the ocean that emerged when the supercontinent Pangaea – all the world's landmasses joined together – began to fracture in the early Mesozoic around two hundred million years ago, opening a vast seaway between Gondwana in the south and Laurasia in the north. The reefs and lagoons of the Tethys teemed with life: ammonites, rudists (strange reef-building bivalves), sea urchins and the ancestors of modern fish, all leaving their traces in the limestones that would one day be raised to form the Apennines and the Alps. The greater part of Italy's story unfolded far to the south, in the bright tropical waters along the margins of Gondwana on crustal blocks that would later become part of the African plate. Here, coral reefs spread across wide continental shelves, their limestone skeletons slowly hardening into rock that would eventually be heaved skyward to become Italy's mountains, hills, quarries and cathedrals.

The Dolomites, Italy's most famous part of the Alps, are perhaps the most spectacular expression anywhere in the world of ancient coral islands turned to mountains. Anyone who has witnessed the *enrosadira* – the roseate alpenglow that bathes these peaks at dawn or dusk – will understand the sense of wonder they inspire. Their pale, craggy summits, chemically distinct from the rest of the Alps through the presence of magnesium in their composition, record the miraculous return of coral reefs in the Triassic after millions of years of eerie absence following the end-Permian catastrophe. This was the greatest extinction event in Earth's history – the 'Great Dying' – which wiped out nearly all life on the planet some 252 million years ago. The Dolomites stand as a monumental record of recovery, showing how marine ecosystems revived far sooner than those on land. And given that our own coral reefs are now predicted to vanish by mid-century, it is oddly consoling to know that they have rebounded before, if only after a very long pause. Even today, the Dolomites' living world reflects their origins as islands. Individual massifs harbour highly individualised enclaves of endemism, from rare alpine plants, such as pink cinquefoils (*Potentilla nitida*) and tufted horned rampions (*Physoplexis comosa*), to narrowly distributed invertebrates, including ground beetles such as *Carabus alpestris dolomitanus,* confined to a single mountain group and nowhere else.

As time went on, through the long ages of the Mesozoic and beyond, some of the world's most beautiful and distinctive building stones were quietly taking shape beneath warm tropical seas and volcanic islands. Long before they were quarried for temples, villas and cathedrals, they lay as sediments on the floor of the Tethys Ocean, the same ancient sea that shaped the Dolomites. Italy's geology was already composing its tricolour palette of red, green and white: a flag written in rock, formed over hundreds of millions of years, mined for the past three millennia and built into the fabric of civilisation.

Just to the south of the Dolomites, Rosso Ammonitico formed in the quiet depths of the Jurassic sea around what is now Verona. This iron-rich limestone, dense with the coiled shells of ammonites, hardened into a deep-red stone rarely matched elsewhere for colour and clarity. Polished to a gleam, it became the warm, blood-coloured marble of the Verona Arena and one of the principal stones of Venice in the Basilica of San Zeno, and punctuating the pavements of St Mark's and the Doge's Palace.

In the western Alps, Liguria and Tuscany, green serpentinite originated deep within Earth's mantle and was later lifted skyward by tectonic collision. As seawater seeped into the rising rocks, it transformed their iron-rich minerals into serpentine, giving Italy its distinctive dark, silky green stone. Its sinuous veining resembles the skin of a serpent, hence its name, and it has been prized since antiquity for columns, altars and inlay. The best-known variety, Verde di Prato, lent its deep lustre to the Florence Baptistery, the Duomo of Siena and the Orvieto Cathedral.

And then there is Carrara, perhaps the purest and most coveted marble in the world. Once a limestone laid down in Jurassic seas, it was later metamorphosed under immense pressure into a stone of almost angelic whiteness: fine-grained, faintly translucent. From it were carved Michelangelo's *Pietà*, Gian Lorenzo Bernini's *Ecstasy of*

St Teresa, the Pantheon's columns and the Arch of Constantine. Today, most of its finest statuary-grade marble is exhausted, yet the quarries still yield some of the world's most sought-after stone, now destined less for sculpted masterpieces than for kitchen countertops.

The Apuan Alps, home to the Carrara quarries, have surrendered more of their precious white rock in the past two decades of mechanised extraction than in the previous two millennia combined. The No Cav – 'No Quarry' – movement, heir to a century of local resistance in Carrara, is among the longest-standing and most symbolically charged environmental struggles in Italy. We have watched, almost indifferently, as an entire mountain range is carved away, its flanks left with gleaming white, terraced wounds too bright to look at in direct sunlight.

Finding Ciro

The majority of Italy's visible rocks are sedimentary: limestones formed from the shells, corals and skeletons of ancient marine creatures; mudstones pressed from fine clays; and sandstones born of vanished beaches and river deltas. These are the rocks that tell the story of Italy's ancient life, preserving the fossilised remains of its earliest residents.

Having lain underwater for most of its geological history, Italy's richest fossils come from the sea. From the earliest traces of underwater life in Sardinia more than five hundred million years ago, through the great Triassic reptiles of Besano on the Swiss border, to the whales and sharks of the Pliocene, they chart the evolution of life beneath the waves like few places on Earth. One site in particular – Monte Bolca, just north of Verona – is of such importance that it illuminates not only the evolution of fish, but the very development of natural history as a science. We shall visit it later.

However, with so much volcanic and tectonic upheaval over the ages, it is unsurprising that Italy lacks anything approaching a continuous record of its ancestral life. Erosion, subduction and repeated uplift have erased or deformed many of the sedimentary layers where fossils are usually preserved. The volatile geophysics of this young land mean that its fossil record speaks only in scattered bursts across hundreds of millions of years.

Such were the gaps that, until as recently as the 1990s, it was thought that dinosaurs – the creatures that dominated Earth for some 150 million years – had never permanently lived on Italian soil. Ironically, Italy's most significant contribution to their story seemed instead to lie in their end. In 1979, at the Bottaccione Gorge near Gubbio, geologist Walter Alvarez identified a thin, iridium-rich clay layer sandwiched between the Scaglia Bianca and Scaglia Rossa limestones – the global fallout from the asteroid impact that ended the Cretaceous, 66 million years ago, and with it the age of the dinosaurs.

Occasional footprints in Tuscany and the Alps hinted at a passing presence, but no bones had ever been found. Then, in November 1980, an amateur fossil hunter named Giovanni Todesco changed that. Newly transferred to work in southern Italy, he visited a quarry above the village of Pietraroja, in the Matese mountains of Campania – one of Italy's few Lagerstätten, unique places where fossils are preserved in exceptional detail. Most were fish entombed in the muds of a warm Cretaceous lagoon, 113 million years old. On that day in 1980, Todesco and his family were about to make one of the most extraordinary and fortuitous palaeontological discoveries of all time.

While exploring the quarry with his wife and two children, Todesco noticed a demolition crew pulverising fossil-bearing rock for road fill. Racing against the digger, he grabbed what slabs he could. Among them was one bearing a faint dark mark, only a few millimetres long. The family left shaken but exhilarated. 'We returned home in silence,' he recalled, 'sad for what had been destroyed, but happy to have saved something, even if we didn't yet know what it was.'[1]

One day later, on 23 November, the same region was devastated by the Irpinia earthquake, one of the most destructive in Italy's modern history. Hundreds of towns and villages were levelled, nearly 3,000 people killed, thousands more injured and around 280,000 left homeless across Campania and Basilicata, with tremors felt as far away as Puglia. Aftershocks and landslides shook the area for months. Amid the devastation, the Todesco family soon returned north to Verona, carrying with them the slabs they had salvaged from Pietraroja, which joined the growing collection in Giovanni's basement laboratory.

Over the next decade, Giovanni and his wife would return to the slabs whenever time allowed, working patiently in their basement after hours. The first fragments, saved by the children from the steamroller, revealed only a scatter of small incomplete fish. It was not until some years later that they turned to the larger blocks Giovanni had rescued from the digger. Matching the slabs together, he began carefully scraping away a centimetre of limestone from above the dark mark he had first noticed in the quarry.

Fully expecting another fish, Giovanni was astonished to uncover instead the stump of a tail, then a pair of hind legs, their bones and cartilage in near perfect preservation. Gradually, more details emerged: the creature's intestines with traces of its last meal, the forelimbs with superb little claws like those of a bird of prey and a long neck rising to a large head with huge baby-bird eyes and a mouth bristling with sharp carnivorous teeth.

It would take hundreds of hours of painstaking work for Giovanni and Giovanna to reveal the full picture: an immaculately preserved, ferocious, little bird-like creature, no more than 23 centimetres in length. Yet they remained baffled by what it might be. They compared their fossil to others in books but never found a perfect match. Their best guess was something midway between a reptile and a bird, perhaps like *Archaeopteryx* (one of the most important fossils ever found and that demonstrates the evolutionary link between dinosaurs and modern birds) but without any sign of feathers.

In 1993, finally conceding that the fossil might be too important to remain in their basement forever, Giovanni contacted Giorgio Teruzzi, curator of vertebrate palaeontology at Milan's Natural History Museum. Teruzzi's reaction was instant and visceral. '*È lui, è lui!*' he exclaimed. 'It's him!'[2] What lay before Professor Teruzzi was not only the first dinosaur fossil ever discovered in Italy, it was also the first dinosaur found anywhere in the world with completely preserved internal organs and soft tissues.

'Sue', the near-complete *Tyrannosaurus rex* unearthed in 1990, certainly took the prize as the blockbuster palaeontological find of the century. But 'Ciro', as Giovanni's fossil quickly became known, was a very close second, and in many ways more unique. Until that moment, palaeontologists had never seen muscles, soft tissues or the digestive system of a dinosaur. The exceptional preservation of Ciro meant that, for the first time, scientists could begin to understand the detailed anatomy of a dinosaur's insides: how muscles were attached to bones, how it stood and walked and even, stunningly, what it had eaten.

Such was the level of detail, with new technology Cristiano Dal Sasso and collaborators at the Natural History Museum of Milan were able to perform an 'autopsy' of the fossil, revealing the complete three-dimensional internal anatomy of this remarkable creature. Ciro was a baby theropod, perhaps only a few days old, still with the remains of his first meal – a small lizard – in his stomach when he died and was miraculously preserved 113 million years ago, for reasons we still cannot fully explain.[3]

After the initial announcement in 1993 and correctly consigning the fossil to the regional authorities in Salerno who govern Pietraroja, five years of intense study followed. In 1998, *Scipionyx samniticus* was unveiled to the world, making the cover of *Nature* and appearing in newspapers across the globe.[4] It was nothing short of a sensation. For the first time in over two centuries of dinosaur discoveries elsewhere, Italy finally had a fossil to make even the great American institutions jealous.

Despite the international acclaim and being at the centre of the greatest moment in modern Italian palaeontology, it marked the beginning of a long nightmare for Giovanni Todesco. He was detained and put on trial for trafficking stolen artefacts, such is the letter of Italian law. What should have been resolved in a matter of hours dragged on for nearly four years before the authorities fully exonerated him, eventually granting a full pardon and rightly declaring him a saviour of Italian cultural heritage. The presiding judge observed, with some irony, that 'the energies spent prosecuting Todesco would have been used more profitably if aimed at preventing, limiting and repressing abuses at the Pietraroja site'.

This overzealous litigiousness points to a deeper structural problem: Italy still applies a single, antiquated legal framework to both archaeological artefacts and palaeontological remains. In practice, this means that fossils are treated like cultural property, subject to the same permits, restrictions and criminal penalties as archaeological treasure, making scientific excavation at times extraordinarily difficult. While there is a widely shared belief that Italy still conceals fossils as extraordinary as Ciro, the laws governing excavation – many of them dating back to the fascist period – make it almost impossible for such finds ever to come to light. Incredibly, since Ciro's discovery there have been no further official excavations at Pietraroja.

Yet perhaps there is room for optimism: the surrounding landscape now falls within the newly established Matese National Park, created in 2025 to protect the mountains and forests of Molise and Campania. With the fossil beds of Pietraroja now inside its boundaries, there is hope that one day the layers that yielded Italy's first dinosaur may once again be explored – this time under the banner of protection rather than prosecution.

Moving On Up

Ciro emerged from a stone quarry that was once a tropical lagoon bed but is now tilted high in the Matese Mountains more than 700 metres above sea level, demonstrating how Italy's restless geology has hoisted ancient seas into the sky.

Yet Ciro himself was a land-dweller, showing that by the Cretaceous, Italy's landscape was already beginning to rise from the depths, its coral-rimmed islands home to small, endemic dinosaurs destined to become the peaks of future mountains. As Africa pressed ever northward, millions of years of compacted seabed were squeezed, folded and thrust skyward to form first the Alps, and later the Apennines. These mountains and hills define the peninsula still, shaping its extraordinary range of climates, ecosystems and ways of life from north to south.

The Alps form Italy's dramatic northern boundary, an imposing natural gateway to continental Europe that has defined the edge of the peninsula for millions of years. Their bedrock was laid down hundreds of millions of years ago beneath ancient seas, began uplifting around 60 million years ago as Africa collided with Europe, and was carved into their present form during the great glaciations of the last 2 million years. At their peak, around 30 million years ago, the Alps may have stood nearly twice their current height, rivaling the modern Himalayas. Since then, they have steadily eroded, fractured and slimmed to almost half their former stature. Much of their current landscape, such as the deep U-shaped valleys, knife-edged ridges and cirques, was sculpted during the last 2.6 million years – the Pleistocene or Ice Age – when immense glaciers smothered the range. Only in the last 12,000 years, as the ice retreated, were the valleys and passes of the modern Alps fully revealed.

Today the Alps harbour a wealth of life: roughly 30,000 animal species and some 13,000 plants. Among the latter are about 4,500 vascular plants and as many as 450 of these are unique to the Alpine arc, from tiny saxifrages clinging to cliffs to dwarf primroses flowering beneath the snow.

Even within single valleys, such as those of the northeastern Limestone Alps, dozens of endemic plants and subspecies have evolved in isolation, each adapted to the thin soils and harsh extremes of the high mountains.

It is geology that drives these differences, shaping soils, slopes and forest cover alike. In the western Alps, granite and gneiss weather into acidic soils that favour spruce, larch and Scots pine; in the east, the dolomitic limestones of the Dolomites, rich in calcium but thinly soiled, host beech and silver fir, giving way higher up to open meadows and dwarf pine.

The Ibex Affair

It is thanks in particular to the creation of Italy's first national park, the Gran Paradiso, that one of the Alps' most emblematic creatures has survived. By the late nineteenth century, the Alpine ibex (*Capra ibex*) had vanished from almost the entire range, hunted relentlessly for its horns, meat and even its supposed healing organs, such as the 'heart cross' – a piece of hardened cartilage from the heart's valves thought to make the bearer invulnerable. Fewer than a hundred animals clung to life in the royal hunting reserve of the Gran Paradiso, protected by King Victor Emmanuel II from 1856 onward – not from hunting itself, but from all hunters save the King. Armed guards patrolled the valleys and poaching was punishable by imprisonment or exile. The ibex survived not through anything we would recognise as conservation today, but thanks to an exclusive royal preserve – and the fact that even kings do not always hit what they aim at.

The Swiss, who had lost their ibex entirely, were desperate to restore it, but the Italian crown refused to release any live animals. So, in the early 1900s, Swiss conservationists turned to covert means: a handful of Aosta Valley poachers were paid to capture ibex kids and smuggle them across the border, disguised as goats in mule trains bound north. Many died during capture or transport, but a few survived and were bred secretly at the Peter and Paul Wildlife Park in St Gallen.

From those survivors the species was reintroduced to Switzerland and, later, to other parts of the Alps. After the First World War, the establishment of the Gran Paradiso National Park in 1922 finally gave the Italian herd lasting protection. What began as a clandestine act of smuggling in a royal hunting reserve became one of Europe's great conservation successes. Every ibex that now roams the Alps, from France to Austria, descends from those few that once clung to the cliffs of the Gran Paradiso.

Spinal Gap

The Apennines are Italy's youngest mountains, born over the past 20 million years and still slowly rising. Though often treated as a distinct mountain-building event (orogeny) from the Alps, they share the same tectonic parentage: the collision of Africa, Eurasia and Adria – the Adriatic microplate, a northern fragment of the African plate that broke away millions of years ago.[5] Around 30 million years ago, the twin blocks of Corsica and Sardinia began to detach from what is now France and Spain. Over the next 15 million years, the block swung anticlockwise into the western Mediterranean like a door pivoting open. Its slow passage triggered a chain of upheavals: Calabria and Sicily were thrust into being as they pressed against Adria, while in Tuscany and the Tyrrhenian basin, deep older rocks were heaved once more toward the surface.[6] Close to where I live in southern Tuscany, one of my absolute favourite wild swimming spots is on a river, the Farma, where some of these 300-million-year-old rocks protrude to form waterfalls and pools.

The movement of Sardinia, Corsica and Calabria set the stage for Italy's backbone. As these blocks rotated into place and ground against the Adriatic microplate, the stress and compression along their boundary began to buckle the seabed. Out of those submarine folds rose the Apennines, the long mountain chain that still defines the peninsula today. The same restless forces that lifted them also opened fractures deep in the crust, allowing magma to rise – the beginnings of the volcanoes whose fire shaped so much of Italy, and which we will encounter in the next chapter.

More than Giuseppe Garibaldi, hero of the Risorgimento, the Apennines are Italy's true unifier. Linking north to south, they stretch from Liguria in the northwest to Reggio Calabria at the southern tip, forming the peninsula's spine, a chain of stone that defines Italy's geography as much as its culture. Including the Peloritani Mountains of northeastern Sicily and Mount Etna further south, the Apennine system stretches for more than 1,500 kilometres. Twelve national parks are anchored in these mountains and within them survives much of Italy's biodiversity, cultural and linguistic diversity, agricultural variety and some of the wildest landscapes left in Europe.* A

* In its long run south, the Apennine chain passes through a polyphony of dialects: Ligurian, Emilian, Tuscan, Umbrian, Romanesco, Neapolitan, Calabrese – each valley and province with its own cadence. Tucked into its folds are even more exotic survivors: Albanian villages in Basilicata and Calabria, Greek-speaking enclaves on the Ionian side, even a pocket of Slavic speech in Molise.

full thirty-six rivers rise from the Apennines, among them the Tiber, Arno, Volturno and Reno. Some thread westward to the Tyrrhenian; others run straight and fast to the Adriatic. These rivers, and the mountains that feed them, continue to shape Italy's geography, politics, culture and economy; they provide the lifeblood to its living landscapes.

Understandably, the Apennines have long frustrated modernisers – Futurists and engineers alike – who dreamed of taming them. For centuries they splintered Italy into lands within lands, forming barriers to unification, isolating valleys and cultures on either side and slowing the flow of armies and commerce. Marauding invaders found themselves stalled and scattered in their folds and valleys, from Hannibal's Carthaginians to German divisions holding the Gothic Line during the Second World War. Time and again, the Apennines turned campaigns into stalemates, created refugees and unlikely victors, and forced history to move at their pace.

The Apennines are the single greatest driver of Italy's terrestrial biodiversity. Their long, fractured chain, stretching the length of the peninsula, has created countless microclimates and refuges where species evolved in isolation, survived glaciations and adapted to every gradient of altitude, soil and rainfall. There are some extraordinary locations, both above and underground. In the northern Apennines, the gypsum formations of Emilia-Romagna give way to underground rivers and halls of translucent crystal. In the far south, the bare limestone peaks of the Pollino massif rise like fossil waves, their flanks studded with thousand-year-old Bosnian pines. And in the Frasassi Caves of Le Marche, entire ecosystems run on sulphur instead of sunlight.

Similar to the Alps, the immense geodiversity of the Apennines – a patchwork of limestones, marls, sandstones and volcanic rocks, each generating its own soils and plant communities – underlies their bewildering floral richness. The limestone ranges of Abruzzo and Molise, fractured and riddled with vast karstic cave networks, support dense beech forests on their cool, humid summits and holm and downy oak on the parched slopes below, with a carpet of montane herbs, crocuses and anemones thriving in between. In the northern Apennines, where sandstones and marls prevail, deeper and more acidic soils favour chestnut and mixed deciduous woods accompanied by slopes rich in heathers, bilberries and other acid-loving shrubs. Scattered among these, ophiolitic outcrops – fragments of ancient ocean crust – create lean, mineral-rich barrens where only a handful of specialised plants endure, such as *Odontarrhena bertolonii* (formally *Alyssum bertolonii*), the so-called 'nickel flower', a northern Apennine endemic capable of surviving on serpentinite soils toxic to most vegetation. Further south, around the extinct volcanoes of Monte Amiata and Roccamonfina, young, fertile volcanic soils once again enrich the land, feeding chestnut groves, orchards and some of Italy's most abundant mixed forests.

From rivers and beech forests to high meadows and caves, the Apennines shelter a hidden nation of endemics: bears and wolves, shrews and toads, black squirrels, Roman moles, salamanders – including my muse the *Salamandrina* – as well as the remarkable lungless cave salamanders and beetles without eyes. Each species endures because the mountains fracture the land into a mosaic of small worlds – valleys and mountains that became islands, and pastures and caves that became sanctuaries. More than any other feature, the Apennines have made Italy the most biologically diverse country in Europe. So many species are confined to these mountains that we will meet them again and again in the chapters ahead, yet still only scratch the surface.

Two species in particular have shaped Italy's story of biodiversity and conservation more than any others. The Marsican brown bear (*Ursus arctos marsicanus*) – one of the world's rarest bears – is an endemic subspecies of the Eurasian brown bear (*Ursus arctos arctos*), once widespread across the peninsula and likely recolonising the Apennines as the Ice Age ended some twelve thousand years ago. As human pressures grew, the population became stranded in the central and southern mountains, and today numbers barely seventy individuals, scattered across the national parks of Abruzzo, Lazio and Molise, the Gran Sasso and the Maiella. It is the poster child of Italian biodiversity, a beloved member of the mountain community, yet also a stark reminder of the limits of the country's wilderness. Despite a multilayered network of organisations working to secure its future – the national parks, Salviamo l'Orso, Rewilding Apennines, WWF Italia and the European Nature Trust – and the national Action Plan for the Protection of the Marsican Brown Bear, its numbers have remained perilously low. Slow to reproduce and hemmed in by roads, villages and shrinking habitat, the Marsican brown bear endures more as a symbol of hope than of recovery – undoubtedly the most emblematic of all Italy's unique and imperilled wildlife.

By the 1970s, the other grey symbol of Apennine conservation – the wolf – had been driven to the brink of extinction in Europe, with perhaps no more than a hundred surviving. All belonged to the endemic Italian subspecies, *Canis lupus italicus,* confined to the most inaccessible forests of the central and southern Apennines. In 1971, biologist Luigi Boitani and conservationist Fulco Pratesi, founder of WWF

Italia, launched L'Operazione San Francesco, named for the saint who preached to wolves, to save the last of them. Their campaign led to full legal protection in 1976, the first such law in Europe. Since then, the wolf has staged one of the greatest comebacks in modern conservation, returning to all parts of the continent. Today, around four thousand roam across the Italian peninsula, while Europe as a whole now supports more than twenty thousand, more than in all of North America. It remains the continent's most remarkable carnivore recovery, achieved without a single reintroduction, driven instead by protection, rural depopulation and the species' own formidable adaptability.

The return of the wolf has also restored a missing balance to Europe's ecosystems. By keeping deer and wild boar in check, wolves have allowed forests and shrublands to regenerate and young trees to take root once more. Their kills feed ravens, eagles, foxes and vultures, while their presence alone reshapes the behaviour of their prey, spreading grazing pressure more evenly across the landscape. In short, the wolf's comeback has revived the ecological pulse of the mountains it once ruled. Yet that success remains fragile: the European Union has recently moved to weaken the wolf's protection status, a shift critics have linked to the killing of President of the European Commission Ursula von der Leyen's pony by a wolf in 2022. It is a reminder of how easily centuries of persecution can resurface, and how quickly progress can be undone.

The northward drift of Africa is the key to understanding how Italy was conceived, carried and born. With few exceptions – Sardinia, Puglia, the Po Valley, parts of Sicily – every corner of the country bears the scars of the orogenic convulsions that raised the Alps and the Apennines. Africa remains the driving force, shaping how the peninsula continues to deform, and how it will one day disappear. Though the familiar boot-shape seems fixed, it is slowly bending eastward, pressed toward the Balkans across the Adriatic. Rising seas will, in the coming centuries, reclaim much of Italy's coast and lowlands, including the Po Valley, returning the peninsula to an archipelago. In the still longer view, the Mediterranean will disappear, as Africa continues to march northward, eventually swallowing Italy and welding itself to Europe.

Italy's mountains may look eternal, yet the ground beneath them is still alive. The same forces that raised the Alps and Apennines continue to push, fold and fracture the crust, storing energy that one day must escape. For all its ancient stones, Italy remains one of the youngest lands in Europe, trembling still on the fault line between creation and destruction.

Out of this tension comes its next story: the story of fire. Where mountains rise, volcanoes are never far behind. The chapter that follows traces that fire as it surfaces in smoke and ash, reshaping the land and, just as often, renewing it.

The bright light of the Calabrian sun was now dimming, the temperature dropping, and I was in need of a beer and some southern cooking. I gathered a few small, flinty grey rocks from the top of the ski area and skimmed them down the mountainside, as day gave way to the long glow of evening.

CHAPTER TWO

Fire

Whenever I look at a mountain I always expect it to turn into a volcano.

Italo Svevo, *The Further Confessions of Zeno*

As football chants come, '*Oh Vesuvio lavali col fuoco*' ('Oh Vesuvius, bathe them with fire') is among the more hostile. Sung by opposition fans, mostly from northern Italy, Neapolitans have long grown used to such inflammatory abuse, but they have nearly always had the last laugh. Their infamous banner aimed at Verona, '*Giulietta è 'na zoccola*' ('Juliet is a whore'), has become immortal in Italian football lore, a Shakespearean taunt to which the Veronesi have never quite found a riposte. Nowadays, Napoli fans have reclaimed the Vesuvius jibes for themselves, belting out '*Vesuvio erutta, tutta Napoli è distrutta*' ('Vesuvius erupts, all of Naples destroyed') to the ubiquitous tune of Gala's 'Freed from Desire'. That brazen celebration of the fiery mountain on their doorstep is admirable, if slightly disturbing, given that the threat of death and destruction on a massive scale is very real.

In 79 CE, Vesuvius erupted with such terrifying speed that it buried Pompeii and Herculaneum within hours, killing several thousand people, none of whom had any inkling that their mountain neighbour was a volcano. It had passed out of living memory that, nearly two thousand years earlier, the same mountain had erupted with equal ferocity, overwhelming a cluster of Bronze Age villages near Nola and blanketing them under metres of pumice – a prehistoric Pompeii whose huts, footprints and even tethered goats were found perfectly preserved in the ash.

Vesuvius has erupted several times since 79 CE, with significant outbreaks recorded in 203, 472 and 512, before falling into a long medieval silence that lasted more than a thousand years. It reawakened violently in 1631 when pyroclastic flows swept down to the sea and destroyed towns along the Bay of Naples, killing thousands. It was then almost continually active, with significant eruptions in 1794, 1906 and 1944 that buried villages and vineyards once again. The volcano's current eighty-year silence is its longest pause since the seventeenth century, but geology suggests it is a respite rather than an ending.

Today, some seven hundred thousand residents live within the *zona rossa*, the highest-risk zone around Vesuvius, and in the case of an eruption, about three million people in the greater Naples area would be exposed to ash and the pyroclastic fallout. Civil protection authorities estimate that a full evacuation would take roughly seventy-two hours to complete – on the optimistic assumption that the volcano provides generous warning. Neapolitans call Vesuvius simply *a muntagna* – the mountain – and accept its presence as they would a temperamental relative: unpredictable, dangerous and beloved.

Nowhere on Earth do so many people live so cheerfully in the shadow of potential annihilation. Naples, with Vesuvius to the east and the Campi Flegrei to the west (the ancient gateway to Hades), is the only metropolis bracketed by two active volcanic systems: one that buried its suburbs in antiquity and another that, some thirty-nine thousand years ago, erupted with such force that it dimmed the sunlight across Europe and scattered ash as far as Russia.[1] Yet across the sea, in miniature, the same story repeats. On the island of Stromboli, barely five hundred people live year-round on the flanks of a volcano that has been erupting almost without pause for two thousand years. Its steady fountains of fire, visible to passing ships since antiquity, have earned it the title 'lighthouse of the Mediterranean'. Here, as in Naples, life continues in the glow of danger, a rhythm so constant that the mountain's heartbeat has become the measure of time itself.

When Roma fans, my own adopted tribe, gleefully join in the anti-Napoli Vesuvius chants, they rarely pause to consider that they, too, live in the shadow of a massive volcano. The Alban Hills, just south of the city and born of the same tectonic unrest that gave rise to Vesuvius and Campi Flegrei, are considered dormant but not extinct. They have been erupting intermittently for the last eight hundred thousand years, with the last major blast occurring about thirty-six thousand years ago, and their explosive debris forms much of the ground on which Rome stands. Today the slopes appear disarmingly peaceful, clothed in vines and umbrella pines, with the crater lakes of Albano and Nemi around which the Castelli Romani villages cluster. At its edge lies Castel Gandolfo, seat of the Papal Palace and the Vatican Observatory. Yet though these 'hills', like Vesuvius to the people of Pompeii and Herculaneum, are easily forgotten as the remnants of a vast volcanic complex, geologists suspect that they once erupted at intervals of roughly thirty to forty thousand years – a period uncomfortably close to the length of time that has already passed.

Volcanologists tend to play down any immediate risk from the Alban hills, yet each year satellite data record a slow uplift of the central caldera – a sign of magma recharge – while fumaroles (vents where volcanic gases escape from the earth) continue to exhale carbon dioxide and sulphur gases. Small earthquakes ripple frequently through the area, reminders that the system is still alive beneath its veneer of vineyards and lakes. Scientists monitor it closely and consider its current calm reassuring, but they acknowledge that, given enough time, its potential for catastrophe rivals that of Naples. Perhaps all fans would be wise to think twice before taunting sleeping giants.

Life Is Tuff

Despite the risks, the real story of Italy's volcanoes is not their latent potential for death and destruction, but how much life they have delivered. Without their fiery presence, both Italy's natural and cultural history would be unrecognisable. They have been the source of the country's energy, biodiversity, agricultural richness, abundant water and the raw materials from which so many of its cities were built. It is almost impossible to imagine an ancient Rome of such power and permanence without the gifts of volcanic stone, fertile soil and mineral-rich waters. And modern Italy continues to reap the unique benefits of living among the only active volcanoes on mainland Europe, even as it learns to coexist with their unpredictable nature.

Italy's volcanoes exist in the first place because the country sits on a restless seam where two continents collide. The African plate is slowly pushing northward, sliding beneath Europe and forcing the crust above it to bend, break and melt. From that friction and pressure, magma is born – molten rock rising from the deep, seeking weak points in the surface.

Over millions of years, this collision lifted the Apennine Mountains and at the same time tore open the Tyrrhenian Sea behind them. While the land on one side was thrust upward into mountains, the crust on the other was drawn thin until it cracked open. Through those fractures, magma reached the surface, giving rise to the chain of volcanoes that still runs down the length of Italy, from Tuscany and Lazio to Naples, Sicily and the Aeolian Islands.

Destruction may be a volcano's first act, but creation nearly always follows. When the dust settles, what remains are gifts of immense value. The first and most obvious of these is stone. When magma emerges and cools, it hardens into volcanic rock – what we often call lava stone. Depending on how quickly it cools, it may form glassy obsidian, porous tuff and pumice or dense black basalt, and much else besides. Collectively, these are known to geologists as igneous rocks – stone born of fire.

Volcanic landscapes yield some of the finest building materials on Earth and, in Italy's case, they have quite literally built civilisation itself. Long before cities, obsidian from the Aeolian Islands was chipped into knives, arrowheads and mirrors – the first gifts of fire to humankind. Millennia later, the Etruscans inherited this volcanic bounty across their heartland of what is today Tuscany and Lazio, where ancient eruptions had buried the land in soft tuffs and harder peperino – stone so abundant and malleable that it became the very fabric of their cities. From these fiery sediments they carved dwellings, necropolises and the great *vie cave* – the wonderful excavated Etruscan roads, which still wind through the mossy, tuffaceous ravines of Sovana, Sorano and Pitigliano, among my favourite walks in Italy. The Etruscans built not just on the land but into it, hewing towns from the living rock in a kind of proto-green architecture that seemed to grow straight out of the earth.

It is no exaggeration to say that volcanoes built Rome. The Eternal City rose from the cooled ashes of one of the largest volcanic regions on Earth – the Roman Magmatic Province (RMP). This immense system comprised four major volcanic districts: Vulsini, Cimini, Sabatini and Albani, all intensely active for most of the last half a million years. Without these volcanoes, it is hard to imagine a Rome of such power and permanence, whose rise came during a fortuitous interval of dormancy, a peace in which its people thrived on the legacy of prehistoric fire. Without this volcanic bequest, the trajectory of Roman civilisation, and perhaps its empire, would have been entirely different.

The famous Seven Hills of Rome – Palatine, Aventine, Capitoline, Quirinal, Viminal, Esquiline and Caelian – are built on thick beds of volcanic tuff, pumice and ash, most of it erupted from the Alban Hills. Augustus may have found a city of bricks and left it clad in marble, but his ancestors found a city already founded on tuff. Light yet strong, porous yet durable, and quarried in vast quantities nearby, tuff (*tufo* in Italian) is one of the most distinctive and historically important volcanic building stones in the world, and Italy offers some of the finest examples of its use across millennia. The Servian Wall, Rome's first great fortification, was built from the warm brown Grotta Oscura tuff, quarried near the Etruscan city of Veii; the Theatre of Marcellus rose on substructures of lionato tuff; and beneath the city, the catacombs and cisterns were carved directly from the same stone.

Life has endured in the tuff of Rome for millennia. During the city's earliest years, one of its first and most consequential feats of engineering was the construction of the Cloaca Maxima. Begun under the Etruscan king Tarquinius Priscus in the sixth century BCE, it started as an open ditch lined with cappellaccio tuff, designed to drain stagnant waters from the marshy valleys between Rome's hills into the Tiber.[2] In time, it was vaulted in more tuff, travertine and peperino, all volcanic stones, and built over, becoming the oldest continuously functioning sewer in the world. This simple but transformative intervention reshaped the landscape, allowing the city to expand and to create directly above it what would become the most famous civic space on Earth: the Forum.

In 1998, scientists from the University of Roma Tre discovered river crabs (*Potamon fluviatile*) living in the Cloaca Maxima, under the Forum. Fittingly nicknamed the 'gladiator crab', these crustaceans grow more slowly, reach larger sizes, live longer and mate at unusual times of year compared with their river-dwelling kin. How they arrived there remains a mystery: they may have been swept in by floods, migrated from the Tiber or descended from ancient river fauna trapped by millennia of engineering. Because of ongoing restoration work, the site is largely inaccessible to researchers, and for now the mystery endures. Yet these crabs stand as symbols of how Italy's wildlife has adapted to thousands of years of human reshaping, finding refuge in the drains built from the most enduring of volcanic stones.[3]

In Hot Waters

Travertine – the great white stone of Rome – is the city's other defining building block and the product of one of volcanism's most magical and enduring legacies: hot springs. Italy has more of these natural thermal sources than any other country in mainland Europe, with over three hundred scattered across its landscape. They arise both from lingering magmatic heat beneath the crust and from the fractures and faults of a restless geology, which open pathways for heated groundwater to rise to the surface. The waters are rich in minerals: sulphur, whose hydrogen sulphide gives the familiar eggy tang, but also magnesium, silica and trace elements long prized for their therapeutic effects.

Some volcanic springs are so rich in calcium carbonate that they turn water into stone. As the hot, mineral-laden water rises and cools, it releases carbon dioxide, causing calcite to precipitate and build up in layers. Over time, these deposits form the pale, banded rock known as travertine, a kind of limestone born from hot eggy water. Its pale terraces still form where such springs emerge, most famously at Saturnia and Bagni San Filippo in southern Tuscany, where warm cascades continue to lay down fresh stone as they fall. The Romans quarried their travertine in abundance at Tivoli, east of Rome, where it is still cut today – the same stone that built their aqueducts, temples and amphitheatres, from the Colosseum to St Peter's.

Today, perhaps the most precious legacy of Italy's volcanic past is energy. In southern Tuscany, I live in the shadow of Monte Amiata, a long-silent volcano that last erupted about three hundred thousand years ago, but whose superheated waters still rise through the crust and are now being put to work. More than thirty hot springs emerge across the region, and to the north, at Larderello, that subterranean heat was first harnessed for power. There, in 1911, the world's first commercial geothermal station began operation, and for the next forty-five years it remained the only industrial-scale plant of its kind.

So with all the freely available building material – tuff, travertine, basalt and much more – what was also necessary was something to bind it all together. In 2013, scientists finally unlocked the secret of why Roman concrete has endured for two millennia, still holding together architectural wonders while modern structures crumble in a fraction of the time. The Romans had discovered that mixing lime with volcanic ash – pozzolana, named after Pozzuoli, a town resting atop the Campi Flegrei, and quarried from the hills around Naples and Rome – produced a mortar that not only resisted cracking but actually grew stronger over the centuries, as new minerals continued to crystallise within it. Recent excavations at Pompeii have now revealed the final piece of the puzzle: Roman builders used a 'hot-mix' method, adding quicklime directly to the mix, creating reactive lime clasts that later allowed cracks to self-heal when water penetrated the stone.[4]

The proof of this miraculous volcanic cement is everywhere: the vault of the Pantheon, still the largest unreinforced concrete dome in the world; the sea walls of ancient harbours at Baiae, Cosa and Caesarea, still hard as rock after two thousand years of waves; and the aqueducts, bridges and cisterns that continue to march across the Italian countryside. By contrast, modern Portland cement often begins to fail within decades and is far more energy-intensive to produce, optimised for speed and uniform strength rather than longevity. Only in very recent years have scientists begun to understand and recreate the chemical processes that gave Roman concrete its extraordinary durability. The Romans, without knowing it, had perfected one of the world's most sustainable building materials, a chemistry born of fire, water and time.[5]

If we take the Colosseum as the crowning glory of Roman architecture, and perhaps the most enduring emblem of the ancient world's greatest empire, we find that it was, quite literally, built by volcanoes. Its foundations and inner walls were constructed from lionato ('lion-coloured') tuff quarried from the Alban Hills; its massive load-bearing piers and facade were faced with the pale travertine of Tivoli, born from volcanic hot springs; its paving stones were cut from dense Alban basalt; and the whole vast structure was bound together with indestructible volcanic concrete – pozzolana.[6] The Flavian Amphitheatre stands, in every sense, as a monument to the marriage of fire and civilisation: a city raised from the ashes of its own geology.

Vita

The Romans had many deities to whom they credited the fertility of their land. Among them, Ceres, goddess of grain; her mother Ops, goddess of plenty; Tellus, Mother Earth herself; and Venus, whose generative powers extended to all living things. Yet the true source of Italy's abundance might be more correctly attributed to Vulcan, the unconventional god of fire. Outcast from birth, the unloved son of Juno and Jupiter took refuge with the Cyclopses in the molten heart of Mount Etna, where he forged weapons for gods and heroes alike. From his subterranean workshop came the heat and minerals that renewed the land. The Romans saw him as a divine smith; we might see him as the spirit of volcanism itself: the hidden artisan who, instead of weapons, actually forged Italy's fertility.

Italy's volcanoes created one of the most fertile landscapes on Earth. When volcanoes erupt, they scatter ash and pulverised rock across the land, a natural fertiliser rich in minerals such as potassium, phosphorus, calcium, iron and magnesium. Over time, rain and weathering break these materials down into fine, mineral-laden soils that retain moisture yet drain well, the ideal conditions for crops. Volcanic ash also carries minerals found nowhere else in such abundance: feldspars, pyroxenes and the green crystals of olivine, which release a steady trickle of nutrients as they weather. Even the glassy particles of ash dissolve into fertile clays that act like sponges, holding both water and nourishment. Out of such dust, Italy's most generous soils were born: resilient, self-renewing and capable of sustaining vines, olives, citrus and grain for millennia.

Today, there are certain fruits, vegetables, nuts and legumes grown around Italy's active volcanoes so distinct that they could not have come from anywhere else. On the slopes of Vesuvius grow both the large, pleated Cuore del Vesuvio tomato and the small Pomodorino del Piennolo del Vesuvio, whose fruits ripen in the sun with almost no irrigation, their flavour so concentrated it borders on the miraculous. Around Mount Etna, the mountain's shifting altitudes and mineral soils bring forth some of the world's most prized citrus, especially the blood oranges flushed with anthocyanins generated by the sharp difference in temperature between night and day. From those same slopes come the vivid green Pistacchio Verde di Bronte, among the most coveted nuts on Earth. On the tiny island of Ustica, an ancient volcano long cooled, farmers still grow lentils so fine and tender they cook in minutes. The capers of Pantelleria set the standard by which all others are measured. On my own local extinct volcano, Monte Amiata, are grown what many consider the best chestnuts in the world: sweet, dense and honey scented. And not far away, on the volcanic plains of Lazio, thrive the spiny artichokes of Albenga and the hazelnuts of the Cimini Hills, both nourished by the same deep, fertile ash that once rained from fire.

When it comes to wine, Italy's regions are usually known for one strength or the other: the great reds of Barolo or Montalcino or the mineral-rich whites of Friuli. But on the slopes of Mount Etna, the alignment of soil, climate, moisture and temperature is so precise that the volcano now produces, in my opinion, a combination of some of the finest red and white wines in the world: Etna Rosso and Etna Bianco, grown across vineyards that range in altitude from 400 to over 1,000 metres.

Alive and Well

When the chemical fertility of volcanic soils is paired with a steady supply of fresh water, farmers are blessed with everything they need to grow food. Yet few recognise how important volcanoes are for the provision of that water, especially during the parched summers of the Mediterranean. They act as natural sponges: their fractured, porous rocks absorb rainfall and snowmelt, storing it deep underground. These hidden reservoirs feed aquifers and springs, releasing water slowly through fissures and faults that thread the volcanic bedrock. Around extinct volcanoes such as those of Lazio and Monte Vulture, entire lakes formed where magma chambers once collapsed and later filled with rain and groundwater. Many of Italy's most dependable water sources owe their existence to volcanoes. In turn, they have become oases of biodiversity, deep lakes ringed by forested hills that harbour species found nowhere else on Earth.

In Basilicata, around the crater lakes of Monticchio near the extinct volcano of Monte Vulture (last active around 125,000 years ago), lives one of Europe's most extraordinary relicts: *Brahmaea europaea*, the European owl moth. Originating millions of years before the Ice Age, it is the ancestor of all modern owl moths: sixteen species in East Asia, one in East Africa and this single survivor in Italy. Discovered in 1963 by Count Federico Hartig, a Trentino nobleman and passionate lepidopterist, the moth has been called a 'living fossil', a creature that by all logic should have vanished but instead miraculously endures, hidden in the forests of an ancient volcano. Its survival in southern Italy has been likened to stumbling upon a pocket of Neanderthals or Homo erectus alive and well in a cave.

The owl moth's presence here is a reminder of how isolation and microclimate can shape survival. When much of Europe froze during the Ice Age, the sheltered slopes of Monte Vulture created a warm, stable refuge where this primitive insect persisted while its relatives elsewhere disappeared. Today, it inhabits barely 50 square kilometres of forest – a final fragment of its former range – among dense stands of ash, privet and phillyrea, the plants to which its life cycle is bound. Each spring, for a single month, adults emerge from their cocoons, their hypnotic brown-and-beige wings flickering in the night as they lay eggs on the manna ash. By May, the caterpillars have fed, pupated and vanished underground, some lying dormant for a year or more before re-emerging, a patient rhythm honed over millions of years. It is the very definition of a relict species.

Its endurance is celebrated at the Grotticelle Reserve close to the lakes of Monticchio, the world's first nature reserve dedicated to a single species of moth. Yet even here, its survival is fragile: the further loss or fragmentation of its already limited habitat would almost certainly spell the end for this small miracle of nature, a creature that has outlived volcanoes, glaciers and empires alike.[7]

But even where volcanoes once gave life, their legacy can still turn deadly. South of Vesuvius lies the Terra dei Fuochi, 'Land of Fires', a name once used to praise Campania's volcanic fertility but now synonymous with one of Europe's darkest environmental crimes. For millennia, this area was known as Campania Felix, 'Fortunate Campania', for its ash-rich soils and aquifers fed by the same volcanic springs that made the land so famously productive. But in the 1980s, criminal networks and complicit industries began dumping and burning toxic waste across the plain between Acerra, Nola and Marigliano, poisoning the very land that once fed an empire. The result is what locals grimly call the Triangle of Death: groundwater and soil among the most contaminated in Europe, cancer rates among the highest and a once-blessed landscape turned against its own people.

In 2025, the European Court of Human Rights condemned Italy for decades of negligence, ruling that the state – having known about the problem since the 1980s – had violated the right to life of the three million residents living within the fallout zone. It was a rare moment of justice, but one that came forty years too late. Over those decades, a landscape created by volcanoes and long famed for its fertility was instead abused and contaminated by the worst forms of human greed and neglect.

Footprints in Fire

As well as giving life and fertility to the land around them, volcanoes also have a tendency to destroy the traces of the past and are not particularly helpful to fossil hunters. Yet long before Pompeii's victims were entombed in ash, an eruption north of Naples preserved an extraordinary exception, a far older encounter between fire and life.

Roccamonfina, the southernmost volcano of the Roman Magmatic Province, was highly active between 650,000 and 50,000 years ago. It had two distinct lives: at first a conical stratovolcano quietly effusing lava and then, about 385,000 years ago, a Plinian monster, blasting ash and debris in convulsions that sent pyroclastic flows racing down its slopes. It was one of those eruptions that created one of the world's most astonishing fossil sites.

Until very recently, residents of the nearby village of Tora e Piccilli recounted the legend of mysterious footprints pressed into a steep wall of tuffaceous rock below the little hamlet of Foresta. They were, it was said, the Ciampate del Diavolo, or the Devil's Footprints – the marks of the only being capable of walking across such ground and leaving his trace. The story likely arose in the nineteenth century, when torrential winter rains scoured away the overlying soils and exposed the haunting prints. To the villagers, their sudden appearance confirmed both the devil's presence and divine retribution through the biblical floods. When sceptical scientists finally examined the site in the early 2000s, however, they revealed something far more extraordinary: not the devil's passage, but the footsteps of early humans – dozens of them, left 350,000 years ago, among the oldest ever found anywhere in the world.

I was myself quite sceptical of these genuinely being ancient, preserved footprints, until I visited the site and saw them with my own eyes. Like boot prints in fresh snow, they are now articulated by generations of lichen growing in and around them, moving diagonally, zigzagging their way down the slope. It's impossible that they can be anything other than human footprints. The site is an extraordinary snapshot of the middle Ice Age, brought to life in a way no skeleton or museum reconstruction could ever achieve. Here, scientists were able to quite literally walk in the footsteps of an early hominin, probably our ancestor *Homo heidelbergensis*, a bridge between *Homo erectus* and the later Neanderthals, *Homo neanderthalensis*. The tracks were pressed into an inclined surface of wet, muddy ash, freshly ejected from the volcano but already cool enough to walk upon. Over time, the peculiar chemistry of the deposit hardened, locking the impressions into stone. A later eruption then buried them in fresh debris, preserving them for hundreds of thousands of years.[8]

Because most of the tracks lead away from the summit, some early speculation imagined their makers fleeing with haste from the mountain of fire, perhaps hooting wildly as they escaped the eruptive fallout. Closer study, however, suggests something far calmer: a group moving slowly, carefully and deliberately. Beside one particularly deep footprint lies the clear impression of a left hand, placed down to steady the walker as they slipped in the soft ash. Remarkably, this was identified only when a modern researcher, retracing the path, instinctively placed their own hand in exactly the same spot for balance. Toes, heels and the arches of the feet are all discernible in some of the tracks, allowing us to reconstruct the movements and gait of these ancient hominins simply by following their path. It is astonishing to think that 350,000 years later, we are attempting to trace our evolution by tracking the footprints of our ancestors, one of the oldest of all human skills.

Scientists have deduced that the tracks were made by a group of hominins, individuals standing perhaps 90–150 centimetres tall – most likely children or adolescents moving with adults in a family group or as a band of youngsters alone, quite possibly up to no good. One trackway has recently been shown to continue onward, linking an existing woodland trail still in use today, now considered the oldest preserved human footpath in the world. Mixed among the human prints are those of other animals: the paw marks of a canid, likely an ancient wolf; the delicate impressions of a deer-like ungulate; and the rounded tracks of a small elephant, probably the straight-tusked *Palaeoloxodon antiquus*, sometimes called the ancient Italian elephant. It could almost be a scene from a prehistoric *Jungle Book*.

Despite the danger of eruption, this was clearly a busy landscape, shared by humans and Pleistocene megafauna alike. The scene invites speculation: why were these early people here amid the warm lava? Were they drawn by food, water or perhaps the lingering heat of the ash? Did they believe themselves far enough from the volcano's fury to take the risk? Whatever their reason, this remarkable fossil site, one of the most astonishing I've visited in Italy, preserves in stone the deep and enduring bond between Italy's volcanoes and its earliest human inhabitants.

Idda

A UNESCO World Heritage Site since 2013, Etna is regarded as a geological marvel – one of the most complex volcanic structures on Earth, with four active summit craters and more than three hundred vents and fissures scattered across its flanks. Ask any volcanologist about Etna and their face will light up: it is both scientifically fascinating and deeply enigmatic, forever surprising residents and researchers alike with its mood swings and unpredictability. Sitting on the doorstep of Catania, Sicily's second city – with which, barring the occasional catastrophe, it has been a largely generous neighbour – Etna has dominated Europe's cultural imagination for millennia, a legendary meeting point of wilderness and planetary energy.

Its origins remain something of a mystery. For all the centuries it has been watched, mapped and measured, geologists still debate why it stands where it does – outside the neat arc of subduction volcanoes in Calabria. What is certain is that it came to life around six hundred thousand years ago, when the area was still a broad marine gulf. Its earliest eruptions left pillow lavas now exposed in the Cyclops Islands off Catania, reminders of its submarine beginnings. Since then, layer upon layer of lava has raised the mountain skyward, until it reached its present 3,369 metres, the tallest active volcano in Europe.

Etna is also one of the richest single biodiversity sites in the Mediterranean. Because of its vast height, from sea level to over 3,300 metres, and the constant renewal of its slopes, it compresses the habitats of an entire continent into one mountain. Within a few hours' walk one can pass from citrus groves and olive terraces to oak and beech forests, and from there to alpine scrub, birch and then bare ash deserts at the summit – an ecological cross-section of Europe in miniature. Every eruption begins the story again: Etna is a self-sustaining laboratory of evolution, a place where lava turns to soil, soil to forest, and forest back to ash. Nowhere else in Europe demonstrates so clearly how destruction and renewal can be the same process.

When an eruption covers a valley in lava, the cycle of recolonisation begins almost at once, though it may take centuries to complete. Much of the flora that returns is found nowhere else on Earth. The first to appear are lichens such as snow lichen (*Stereocaulon vesuvianum*), pale grey smudges that begin to weather the new rock within a decade. Then come the first herbs and grasses, notably the hardy dock of Etna (*Rumex aetnensis*), which roots itself in the thin, unstable ash. Cushion plants like the pink Sicilian soapwort (*Saponaria sicula*) and thorny Sicilian milkvetch (*Astragalus siculus*) follow, anchoring the loose sand and sheltering insects in their spiny domes. Over time, the bright yellow Etna broom (*Genista aetnensis*), takes hold – a nitrogen-fixing shrub that creates the first true soil where none existed. From these young soils rise the pale trunks of the Etna birch (*Betula aetnensis*), descended from northern forests marooned here after the last Ice Age – the only birch south of the Alps, and a ghost of colder worlds long vanished. Higher still, above the last trees, grows the Etna violet (*Viola aethnensis*), a tiny alpine survivor that clings to scoria and cinders near the summit, flowering between snow and steam. Here, fire and life are in perpetual dialogue: every eruption a reset, every plant a beginning.

Mount Etna is a living island of evolution, home to an astonishing range of creatures that have adapted to its ever-changing slopes. Among its most distinctive fauna are the Sicilian rock partridge (*Alectoris graeca whitakeri*), the vividly green Sicilian wall lizard (*Podarcis waglerianus*), the elusive Etna weasel (a local form of the least weasel, *Mustela nivalis*, sometimes described as *Mustela nivalis aetnensis*) and the delicate Etna aurora butterfly (*Anthocharis damone*), whose orange-and-white wings drift between broom and birch. Reptiles such as the leopard snake (*Zamenis situla*) and the Etna viper (*Vipera aspis*) warm themselves on black stones, while birds of prey – buzzards, kestrels, peregrines and golden eagles – patrol the skies. Invertebrates have diversified here as though on an island, with beetles, spiders and butterflies found nowhere else on Earth. Higher up, foxes, wildcats and hares climb through birch woods that tremble with wind and heat, completing the most improbable of ecosystems – one that has learned not just to survive fire, but to depend on it.

Walk the villages on Etna's slopes and you find the roads and courtyards coated in a gritty black sand: tephra, the solidified shards of lava and rock that settle after eruptions. Etna's basaltic magma produces dark ash rich in pyroxene, olivine and plagioclase – almost everything plants could ask for. If the prayers of farmers for fertile soils were ever answered, it would be by this strange rain from the heavens. For the ordinary citizen, of course, it is more nuisance than blessing, piling up in drifts that need clearing in a part of the world where maintenance is not always done with the greatest diligence.

Both volcanologists and local residents regard Etna as a living being, sometimes called Idda, 'she' in local dialect, but most often simply as *a muntagna* – the mountain – a female force whose moods and rhythms shape life on its slopes. Boris Behncke, a volcanologist at the Istituto Nazionale di Geofisica e Vulcanologia who has lived with Etna for more than thirty-five years, speaks of the mountain as though of an old friend whose temperament has changed. Over dinner in Milo with his wife, Caterine, a seasoned local guide, he described how the volcano's traditional effusive flows have in recent years given way to more explosive paroxysms, often with little or no warning. Another guide, Paolo Perrisi, told me of an eruption he had reported to the monitoring station in Catania that had shown no clear precursors on their instruments at all. Etna's magma rises from depths we cannot see, its chemistry dictated by forces beyond our reach; in that sense Idda truly is alive, a shape-shifting giant who stirs, rages and then drifts back into uncertain sleep.[9] There was wistfulness in Boris's voice: admiration for Etna's defiance of geological rules, but also unease at how its behaviour is shifting, posing new risks for the surrounding communities.

The next morning, with Etna smoking gently above us, I took my wife and daughters up the Mareneve road to Piano Provenzana. We skied the Coccinelle pistes, edged by black lava stones, in conditions that felt almost Alpine, gazing out across the bay of Taormina. By lunchtime, I'd had a dip in the sea and was eating pasta on the beach, before heading back to Tuscany that evening. I am not sure where else in the world one can ski a volcano in the morning and swim in the sea the same day.

Bad Vibes

It is important to remember that, despite their destructive potential, volcanoes are overwhelmingly life-giving, and in Italy they have provided much of the country's natural wealth. The same cannot be said for the other force born of colliding plates: earthquakes. Italy lies at the collision point of the African and Eurasian plates, complicated by the smaller Adriatic microplate caught between them. Their grinding, subduction and slow shifts build immense stresses in the crust, which are suddenly released when rocks snap or slide along a fault. That release radiates outward as seismic waves, what we feel as an earthquake. The Apennines, Calabria and Sicily are especially vulnerable, their restless geology marked by subduction zones and mountain belts that are still being stretched and thrust upward. It is the unfortunate flip side of living in some of the most beautiful landscapes in the world. These are Italy's most familiar disaster – sudden, frequent and deadly – and fatalities from them far exceed those from volcanoes. Italians know the grim rhythm of devastation: Amatrice in 2016, L'Aquila in 2009, Irpinia in 1980, Friuli in 1976, Avezzano in 1915. Each name carries its own tragic weight.

But none compare to Messina. At 5.20 on the morning of 28 December 1908, an earthquake of moment magnitude 7.1 (equivalent to 7.5 on the Richter scale) struck offshore in the narrow strait between Sicily and Calabria. For 30 or 40 seconds, the two great cities on either shore were shaken to pieces: 95 per cent of Messina and 80 per cent of Reggio Calabria collapsed, trapping thousands beneath the rubble. Submarine landslides triggered a tsunami that sent walls of water up to 13 metres high surging over the coasts, sweeping away entire neighbourhoods within minutes. Aftershocks rattled the region for weeks, finishing off weakened buildings and crippling rescue efforts. Between 75,000 and 120,000 people died – the deadliest earthquake in Europe's recorded history. The scar it left was more than physical: it prompted mass migration from southern Italy, many crossing the Atlantic to begin new lives in America. In the century since, the strait here has remained a place of both separation and aspiration, with successive governments proposing, postponing and reviving plans for a bridge across one of Europe's most unstable fault zones.

Considering all her volcanic restlessness and ground-shaking moods, I have come to think of Italy as a pimply-faced teenager: bursting with energy yet prone to sudden sulks, grumbling discontent and the occasional fiery outburst. She suffers from growing pains, lumpy intrusions and hot effusions, fertile flushes and unpredictable tempers. She cannot yet break free from her continental parents, Africa and Eurasia, who hold her in place and keep her under constant strain, giving her an unstable footing.

But compared with the rest of Europe, which is largely middle-aged or well into its geological old age, we are fortunate to witness Italy in this tumultuous adolescence: full of beauty, confusion and the raw vitality of youth. She teaches us endlessly about Earth's dynamics, even as she threatens to scold us. We should cherish her, but we taunt her at our peril.

CHAPTER THREE

Water

Water is the driving force of all nature.

Leonardo da Vinci, *Codex Leicester*

In a secluded limestone valley near the borders of Lazio and Abruzzo lies one of the strangest and most beautiful lakes in Italy: Posta Fibreno. Its waters are incredibly clear, fed by a network of karst springs, underground channels carved through the limestone, that rise straight from the Apennine depths. Even in the heat of July, they remain cold to the touch. Life here is quiet but abundant. The shore is edged with willow, poplar and reed, shading clear shallows thick with submerged plants. Dragonflies zip over the water and herons wade along the margins, while European pond turtles surface onto fallen branches to warm themselves. In the colder months, the lake becomes a stopover for migrating waterfowl; in summer, reed warblers and coots animate the fringes with their calls.

Beneath the surface lives the lake's most extraordinary inhabitant: the Fibreno trout (*Salmo fibreni*), a fish found nowhere else on Earth, and a survivor of a far older Mediterranean world. It moves like a flicker of bronze through the cold springs and channels, small and secretive, spending its entire life within these few hectares of crystalline water. Unlike other trout, it rarely ventures into rivers or streams; its world begins and ends here. For centuries, local fishermen spoke of a 'little salmon' that never left the lake, unaware that it was a relict species distinct from all others. Genetic studies have since confirmed it as one of the most ancient and isolated salmonids in Europe, a lineage that has survived, unchanged, through cycles of ice and drought, fire and flood, for several million years. It likely owes its isolation to one of the most extraordinary events in Italy's geological history.

Around six million years ago, with something close to the modern outline of Italy in place, the Mediterranean faced one of the greatest crises in its history. Africa, inching northward, sealed the Atlantic gateway at Gibraltar. Cut off from the ocean, the Mediterranean was left to evaporate beneath a relentless sun. For nearly half a million years, seawater vanished faster than rivers could replenish it, leaving behind a vast basin of salt flats and hypersaline lakes where only extremophiles could endure. The islands and volcanoes of the Mediterranean rose like mountains from a blinding, salt-white desert, where ground temperatures in the deepest depressions may have exceeded 60°C. The Rhône, Nile and Danube cut deep canyons thousands of metres below their present beds, while Italy's rivers, stranded in isolation, began to evolve their own distinct freshwater faunas, a legacy that still endures in places like Lake Posta Fibreno.

This event, known as the Messinian Salinity Crisis, reshaped the entire Mediterranean world. When drilling cores from the seabed in 1970 revealed salt layers more than 2 kilometres thick, the scale of the catastrophe became clear: the sea had truly disappeared. Eventually, the Atlantic returned with apocalyptic force. When the land bridge at Gibraltar eventually collapsed, a wall of water poured back into the basin, carving deep scars into the seabed and refilling the Mediterranean in what may have been the greatest flood in Earth's history: the Zanclean Deluge.

For Italy, the legacy of that vanished sea is written in salt and in life itself. When the Mediterranean dried, its rivers and springs were marooned in isolation, becoming evolutionary islands where ancient lineages like the Fibreno trout endured. The Messinian Salinity Crisis helped create a patchwork of freshwater pockets across the proto-Italian peninsula, each a miniature laboratory of adaptation and survival. The Fibreno trout (*Salmo fibreni*), the Italian brook lamprey (*Lampetra zanandreai*), the Tiber barb (*Barbus tyberinus*) and the Italian nase (*Chondrostoma soetta*) are among the heirs to this long history of isolation – living relics of a lost inland world, perfectly tuned to the cold, spring-fed refuges that still thread through the Alps and Apennines.

Fortunately for Italy, even then its waters and their inhabitants were sustained by what may be the most important feature of the landscape: its vast underground reservoirs. Few visitors to Posta Fibreno realise they are standing above a system that runs for many kilometres through the Apennine foothills – the veins of an ancient mountain still pulsing with fresh water.

Inferi

Beneath Italy lies another country entirely: a hidden world carved by water. Over millions of years, rain and rivers have dissolved the peninsula's soft limestone, hollowing out mountains from within to create one of Europe's most extensive karst landscapes. Nearly a fifth of Italy consists of soluble rock, riddled with fissures, shafts and underground rivers that vanish for kilometres before re-emerging at the surface. The uplift of the Alps and Apennines not only raised Italy's peaks but fractured the bedrock beneath them, opening the way for water to sculpt an immense subterranean world.

Today, more than 30,000 natural caves are documented across the country, with new ones discovered each year by some 3,000 cavers working through 170 speleological associations. A few are spectacular enough to be open to the public – around 60 can be visited as show caves, among them the Grotta Gigante near Trieste, whose central cavern is an awe-inspiring 98 metres high, 76 metres wide and 167 metres long. Some of the most extraordinary I've visited are the Frasassi Caves in Le Marche, where entire microbial communities survive in darkness by feeding on sulphur and other chemical energy rather than sunlight, a self-contained ecosystem that mirrors the conditions of deep-sea vents.

The vast majority of Italy's caves, however, remain hidden, carrying rivers and entire ecosystems through the silent underworld. This buried network acts as a huge natural reservoir, absorbing rainfall and snowmelt and releasing it slowly through thousands of springs. These are the secret lungs of the land – absorbing, filtering and breathing out the water that sustains life above. It is estimated that Italy's underworld holds tens of billions of cubic metres of groundwater, volumes comparable to Italy's largest lakes. These colossal underground reserves sustain rivers, replenish lakes and nurture wetlands and have long underpinned human settlement across the peninsula. In Sicily, however, years of drought and the exploitation of the island's shallow aquifers have pushed attention deeper into the underworld in search of water security. Far below the ancient Hyblaean Mountains in the southeast of the island, scientists believe there may be vast groundwater bodies trapped within rocks laid down during the Messinian Salinity Crisis, when sea levels in the eastern Mediterranean fell by as much as 2,000 metres.[1] Modern drilling technology may now be able to reach these reserves, though many see such efforts as a last resort – mistaking a geological inheritance for a renewable supply. Like a bank account from which only withdrawals are made, and never deposits, the exploitation of Italy's deep groundwater raises worrying questions that are likely to recur in the uncertain years ahead.

Beyond the precious gift of water, Italy's massive underworld has also given rise to creatures that have never seen the light of day. In these lightless realms, evolution has written an entirely separate chapter of Italian biodiversity: pale crustaceans, blind salamanders alongside a few highly specialised fish have adapted to a world of perpetual darkness, navigating by touch and vibration. In the dark aquifers of Puglia survives a relic of the vanished sea: *Typhlocaris salentina*, a blind, glassy crustacean known only from a single cave near Capo di Leuca. Descended from marine ancestors trapped underground when the Mediterranean evaporated six million years ago, it has endured ever since in a pocket of brackish groundwater, a living relic of the ancient ocean that once covered Italy. In Piedmont, researchers have recently described a new species of subterranean amphipod, *Niphargus rotundus*, a tiny, blind crustacean never seeing daylight yet thriving in the hidden waters beneath the hills.

Still more remarkable is the olm (*Proteus anguinus*), a lungless, eyeless salamander – Europe's only truly subterranean vertebrate – that can live for more than a century in total darkness. It reaches maturity only after fifteen years, may go a decade without food and has scarcely changed in form for tens of millions of years. Retaining its larval gills throughout life, it breathes and hunts in the flooded caves of the northeastern karst, perfectly adapted to a world without sunlight. During periods of heavy rainfall, researchers have even found olms washed far out into the Adriatic Sea.

New Devilry

The presence of these ancient relict species reminds us how recently Italy emerged from the water. Creatures like the olm and *Typhlocaris* are the last heirs of an Italy that once lay almost entirely beneath the sea. The story of Italy's water begins not with rivers and lakes, but with the sea that made them.

About five centuries ago, hunters and woodsmen stumbled upon a hilltop north of Verona and made a discovery that would transform humanity's understanding of the world. Scattered through the forest floor, they found pale grey stones that cracked open to reveal the imprints of fish and other sea creatures. Believing this to be some form of devilry, they consulted the local priests, who reassured them that the fossils were relics of Noah's Flood: fish stranded as the waters receded, their bodies turned to stone. Reassured, the villagers began to collect and trade them, marvelling at nature's strange handiwork.

The earliest written record comes from 1554, when the physician Pietro Andrea Mattioli described 'stone slabs brought from the Veronese area, in which, by opening them in the centre, various species of fish are found carved in stone… so great and marvellous are the works of nature'. Within decades, the Pesciara of Bolca became the world's first dedicated fossil quarry, a mine devoted entirely to unearthing life from deep time. Quarrymen split the finely laminated limestone like pages in a book, each revealing exquisitely preserved fish – angelfish, morays, rays and wrasse, instantly recognisable to any modern snorkeller, though more akin to those of the Indo-Pacific than the Mediterranean.

At first, they were presumed to be only a few thousand years old – remnants of Noah's Flood or curious 'sports of nature' impressed in stone. As the centuries passed and understanding deepened, the age of the Bolca fossils was pushed ever further back, until finally, in the twentieth century, their true antiquity was revealed: some fifty million years, from a vanished tropical sea that once covered northern Italy.

Over time, more than 250 species and some 100,000 individual fossils have been recovered from Monte Bolca, many still bearing traces of colour, pattern, even the ghostly outlines of flesh. Together they record one of the most complete tropical coral reef ecosystems ever discovered, preserving some of the earliest known fossil representatives of modern reef fish. It dates from that distant epoch when Italy lay beneath the warm shallows of the ancient Tethys Ocean. Here, nature preserved the recovery of life after the extinction of the dinosaurs with almost miraculous fidelity. The mechanism of preservation remains debated, but many scientists believe the site records a series of sudden die-offs in coastal lagoons periodically choked by algal blooms. The dead fish drifted down into anoxic muds, sealed by fine carbonate sediments and perhaps by the fallout of nearby volcanic eruptions. Much later, the slow heaving of the earth's crust during the Alpine orogeny lifted this tropical seabed to 800 metres above sea level, turning a coral reef into a mountain.

From the Renaissance onward, Bolca became a shrine for Europe's natural philosophers. Ulisse Aldrovandi gave its fossils pride of place in his cabinet of curiosities; Giovanni Arduino, studying the strata around it, proposed one of the first geological timescales; and in 1800 Giovanni Serafino Volta published *Ittiolitologia Veronese del Museo Bozziano*, the first scientific monograph on fossil fish. Thanks to Bolca, the birth of palaeontology took on a distinctly Italian accent.

As well as being among the most varied and complete fossils ever discovered, Bolca specimens also became the first true collectors' items of the natural world – early icons of modern palaeontology. Highly prized, the finest examples commanded substantial sums, and such was their abundance that today Bolca fossils can be found in almost every major natural history collection worldwide.

Today, the Cerrato family are still the only holders of the concession to quarry fossils from the Pesciara, having passed their palaeontological craft through successive generations. Visitors can even split the stone themselves and keep whatever fifty-million-year-old treasures they reveal – a privilege unique in Italy, though one that may soon end if Bolca achieves its long-sought UNESCO World Heritage status.

Among all of Italy's natural wonders, Bolca stands alone: a perfect fossil reef, the most vivid window anywhere on Earth into the rebirth of marine life after catastrophe. Its stones remind us that the sea once covered this land, and that life, when entombed, sometimes finds its truest immortality. Bolca fossils are striking objects even in Italy, a country overflowing with wonders. They stand as one of nature's best retorts to anything the great Italian artists ever tried to conjure. Sadly, Leonardo da Vinci missed seeing the fossils of Monte Bolca by a whisker, but had he done so I suspect he would have died satisfied that his creed 'nature alone is the master of true genius' was proven true. Little did people know that the great man had already had his epiphany about the age of Earth, thanks to a different set of miraculous Italian fossils.[2]

The Whale on the Hill

Around the same time that the fishbowl of Bolca teemed with tropical life, a remarkable transformation was taking place in the world's coastal shallows. Certain mammals – land-dwellers resembling small hippos – began spending more time in the water than on land. Their legs, once built for walking, became paddles; their bodies grew sleek and buoyant. Over generations, these pioneers of the shore evolved into the first whales and dolphins. On the opposite side of the Tethys, another group of mammals, ancestors of elephants, waded in and stayed, giving rise to the sirenians: dugongs and manatees, the sea's first fully aquatic herbivores.

Italy, still largely submerged beneath the warm waters of the late Tethys, lay exactly between these two evolutionary experiments – the meeting point of carnivorous and herbivorous lineages that would one day define the Mediterranean's great mammals. The fossil record that followed is one of the richest anywhere on Earth – almost as if the whales themselves had chosen these hills to die on. From Piedmont, Tuscany and Emilia-Romagna to Puglia and Sicily, hundreds of fossil whales have since surfaced from hills that were once the sea floor, their bones lifted high by the slow heave of the Apennines.

Nowhere else in Europe preserves such a continuous archive of marine mammal evolution. The hills of Asti in Piedmont, once a shallow gulf known as the Gulf of Whales, have yielded more than forty complete skeletons, from newborn calves to adults scarred by shark teeth. They record a Pliocene sea, five million years ago, when Italy's coasts teemed with life in the wake of the Messinian Salinity Crisis. Among the remarkable cetaceans found here is *Archaebalaenoptera castriarquati*, the world's earliest known ancestor of the great rorquals (including blue and fin whales) whose discovery helped rewrite the story of baleen whale evolution. Further south in Tuscany was unearthed *Orcinus citoniensis*, the oldest known relative of the modern orca, while in the limestone of Lecce the 8-metre *Zygophyseter varolai* – the formidable 'killer sperm whale' and early relative of today's squid-eating sperm whales – once prowled. Its massive head housed the spermaceti organ, a specialised structure used to focus sound for echolocation, while its powerful jaws carried rows of huge conical teeth. In life, *Zygophyseter* looked oddly like a carnivorous Airbus Beluga cargo plane, its bulbous outsized forehead an adaptation for carrying sonar rather than freight.

Together, these fossils reveal an unbroken chain from primitive Eocene swimmers to modern cetaceans. And the story is not only ancient. The same Ligurian and Tyrrhenian basins that once entombed these leviathans now host eight living species of whales and dolphins. The Pelagos Sanctuary, the world's first international marine reserve for cetaceans, lies over their bones. Here, fin whales, sperm whales and striped dolphins still surface each summer, heirs to a lineage that began fifty million years ago in the waters of the Tethys.[3]

Long before Italy's great cetacean discoveries, Leonardo da Vinci may have encountered a fossil whale himself. In the *Codex Arundel* (c.1480–1518), written when he was twenty-eight, he describes entering a cavern near his home in Tuscany, often identified as the Cava di Spicchio at Collegonzi, and finding the bones of a giant sea creature embedded in the rock:

> *Having wandered for some distance among overhanging rocks, I came to the entrance of a great cavern … [and after some hesitation I entered] drawn by a desire to see whether there might be any marvelous thing within … O Time, swift despoiler of created things, how many kings, how many peoples have you undone? How many changes of state and circumstances have followed since the wondrous form of this fish died here in this winding and cavernous recess? Now unmade by time you lie patiently in this closed place with bones stripped and bare, serving as an armature for the mountain placed over you.*[4]

Whether Leonardo saw a real fossil or wrote metaphorically hardly matters. The passage probably marks the moment he recognised that fossils were the remains of once-living creatures, entombed layer by layer over unimaginable spans of time. Three centuries before Charles Lyell first codified modern geology, Leonardo's notebooks outlined the basic principles of stratigraphy and palaeontology, quietly refuting the notion that they were relics of Noah's Flood. 'Such an opinion,' he wrote of that belief, 'cannot exist in brains of any extensive powers of reasoning.'

With the ever-present risk of being accused of heresy, Leonardo – ever the pragmatist – wisely kept these views to himself. It would take centuries before the true significance of the observations buried in his notebooks came to light. Yet his insight that the mountains of Italy were built from the bones of the sea makes him one of the true fathers of palaeontology. Every fossilised whale and fish unearthed since confirms what he first grasped in that Tuscan cave: that water is the great historian of Earth, recording in stone the rise and fall of life itself. All the more reason to celebrate the magnificent whale fossils found in Italy, not only as colossal marvels of evolution, but as enduring symbols of humanity's awakening to the depths of geological time.[5]

The Waters Above

By around 5.3 million years ago, when the Mediterranean finally refilled, life returned swiftly, recolonising the bare shores and newly flooded shallows. The world had entered a new chapter: the Pliocene. Temperatures were two or three degrees warmer than today, but Earth was already on course for the coming Ice Age. It was during this time that Italy took on something close to its modern outline, though fluctuating sea levels continued to redraw its edges. The Apennines were still rising, the Tyrrhenian still widening to the west, and in the north an immense trough stretched between the newborn mountains.

What had long been a shallow gulf now began its slow conversion into land. After winding unseen through caverns and aquifers below, Italy's waters finally rose to the light. Rain gathered on the slopes of the Alps and Apennines, spilling down in more than a thousand rivers and streams – about 1,200 in all – that web the land from summit to sea. Italy's rivers reflect the restless, mountainous topography of the country: most of them are short, steep and swift. On the eastern flank of the Apennines, they run in straight courses toward the Adriatic, following the mountain folds; on the western side, they wander more erratically, diverted by ancient volcanic uplifts and the slow stretching of the Tyrrhenian margin.

Few of Italy's rivers ever grow large enough to be continually navigable, and none rival the broad arterial waterways of northern Europe. South of the Po basin they behave more like torrents than highways, flashing with winter floods and shrinking to ribbons of silver in the summer heat. Where they reach the plains or the sea, they often slow and merge, leaving much of Italy's limited lowlands naturally waterlogged. Rising sea levels since the end of the last Ice Age only deepened this inheritance, creating fringes of marshes, lagoons and deltas where land and sea continue their long negotiation. Until only a few centuries ago, as much as 10 per cent of Italy's surface was covered in wetlands. Much of the country's modern history has been devoted to their erasure – draining, reclaiming and domesticating the land.

Through the centre of Italy's vast lowland – the Po Valley (Pianura Padana) – flows its great river, the Po. Fed by Alpine meltwater and a web of tributaries, it runs for 652 kilometres, binding together the cities of Turin, Piacenza, Pavia, Cremona, Mantova and Ferrara before dissolving into the Po Delta, one of Europe's richest wetlands. Here, in the backwaters and flooded forests, survive some of Italy's most remarkable freshwater endemic fish: the ancient Italian brook lamprey (*Lampetra zanandreai*) and the last population of the Adriatic sturgeon (*Acipenser naccarii*). Each is a survivor from an older world: the lamprey, a jawless primitive little changed since the Devonian, and the sturgeon, the river's armoured monarch, whose lineage reaches back more than two hundred million years.

If you think your life sucks, imagine being an Italian brook lamprey. This ancient, eel-like creature has followed the same miserable, but ultimately successful, blueprint for more than 360 million years: a living echo of the first vertebrates to make their home in freshwater. It has no jaws, no scales and makes do with a soft, circular mouth fringed with tiny teeth, built not for biting but for hanging on in quiet desperation: filtering, sucking and surviving. For most of its five-to-seven-year existence, it remains in the larval stage, buried in the gravel of clear rivers, quietly hoovering up detritus and silt. Then, in a brief, final burst of purpose, it transforms: its body reshapes for one short season of spawning. That's when the good times end; once adult, its intestine degenerates over the course of six months, leaving it unable to feed. It breeds, scatters its eggs in the gravel and starves to death. Quite possibly the least glamorous life cycle in Italy, though it gets the job done.

The Adriatic sturgeon, by contrast, may live for half a century or more, as if to remind the lamprey what evolution might have offered if it had just held out a little longer. The last native giant of the Po, it is sometimes called a living fossil, possibly just to deliberately annoy palaeontologists, and has changed little since the age of the dinosaurs. Its body is clad in bony plates and its mouth is a soft, whiskered vacuum built for trawling riverbeds. Once common throughout the northern Adriatic basin, from the Po and Adige to the Tagliamento and down the Dalmatian coast, it was fished for centuries for its prized flesh and caviar. Roman gourmands already praised it; by the Renaissance, the dukes of Ferrara served it as the centrepiece of banquets, hauling the immense fish upriver on carts wreathed with flowers.

By the twentieth century, overfishing, dams and pollution had driven the Adriatic sturgeon to the brink of extinction. Today, only a few surviving adults and restocked juveniles linger in the lower Po and its tributaries, their fate now entirely dependent on human stewardship. Italian and Slovenian biologists are collaborating to restore breeding populations, releasing hatchlings into the same gravel beds where their ancestors spawned long before cities rose on these plains. Beneath the mud of the Po Delta swims a lineage that first tasted saltwater before there was even an Italy: a living emblem of continuity between the ancient Tethys and the modern Mediterranean world.

These ancient fish are reminders that Italy's rivers still flow with the memory of the sea, and that their loss mirrors the changing climate of Europe itself. In colder ages, salmon once climbed the Adige, the Brenta and even the upper Po to spawn in the same gravel beds now buried beneath the plain. Their disappearance marks the warming of a continent: as the glaciers withdrew, so too did the great migratory fish, leaving only trout and eels to haunt Italy's waters. The last wild salmon was recorded in the nineteenth century. Today, its bones lie entombed in the Pleistocene muds north of the Po – ghosts of colder rivers, haunting a land that has forgotten its ice.

In one of history's quiet ironies, Italy has become one of the world's largest importers of farmed salmon, almost all of it from Norway, to satisfy a national passion for sushi. Milan alone now ranks among the leading cities worldwide for sushi restaurants per capita, a curious twist in the country's long relationship with fish. The rivers have lost their salmon, but the appetite for them has never been greater. At least the rice beneath it is probably homegrown: Italy is the largest producer of rice outside Asia.

Reverence

Italy's relationship with fish may have become globalised – Norwegian salmon by way of Tokyo – but the country's true freshwater aristocracy is still homegrown and, frankly, underappreciated. They don't come on ice or in rolls; they crawl, croak and occasionally look like they've seen the end of the world. The amphibians have been here since long before the Romans, and, if luck holds, they might just outlast us too.

Each spring they rise from the mud and leaf litter – descendants of ancestors that first emerged in the Carboniferous, three hundred million years ago. The Italian agile frog (*Rana latastei*) calls first, chuckling through the floodplain woods of the north while mist still hangs between the alders. It breeds early, seizing the brief window when meltwater pools fill the hollows of the Po Valley before vanishing by summer, a living reminder of the vast wetlands that once covered a tenth of the peninsula. Because it depends on clean, shaded water and ephemeral ponds, it has vanished from most of its former range, surviving only in fragments of the Po basin – in the Ticino Valley, the Bosco della Fontana near Mantua, and a handful of other protected refuges. A relic of the postglacial floodplains, it endures as a fugitive of its own element – one of the last voices of Italy's ancient freshwater world.

In the Apennines lives the Apennine yellow-bellied toad (*Bombina pachypus*), one of Europe's oldest surviving amphibians. A member of the ancient Bombinatoridae family, kin to the midwife toads and painted frogs, it has changed little since the time of the dinosaurs. When threatened, it advertises its toxicity with a defiant flash of blue and orange, flipping onto its back in a gesture both theatrical and futile. Once common from Liguria to Calabria, it has disappeared from much of its range, undone by disease and the loss of shallow pools. It thrives best in damp mountain pastures where grazing cattle and sheep create puddles and hoofprints that hold water – a patchwork of accidental habitats in which it breeds and hides. To encounter one today in a mountain stream, or more often in a mossy drinking trough, feels almost like meeting a truly ancient amphibian survivor, a creature whose lineage stretches back far further than our own, and one deserving of much greater reverence. Across the Tyrrhenian, its cousins persist, the Tyrrhenian and Corsican painted frogs (*Discoglossus sardus* and *Discoglossus montalentii*), species just as ancient, still calling from rain-fed pools in a landscape that remembers the sea from which they came.

And then there is the spectacled salamander (*Salamandrina*), Italy's own little miracle, and the first spark for writing this book. Endemic to the Apennines, it belongs to one of the most ancient amphibian lineages in Europe – a deeply isolated branch of the salamander family tree and the only living genus of its subfamily found exclusively in Italy. Barely 10 centimetres long, dark brown to black above and striking beneath – a black belly marbled with pale markings, its limbs and tail flushed red to orange – with pale arcs over its eyes that give it a look of perpetual astonishment, it lives hidden in the moss and leaf mould of Italy's cool, shaded woodlands, especially beech forests. When disturbed, it arches its tail and raises its limbs in a slow, balletic display – part warning, part ancient ritual. Its tiny heartbeat still keeps time with the ancient rhythms of the Apennine woods, unchanged for millions of years.

South, in the sun-baked wetlands of Sicily, lives another survivor of Italy's ancient waters: the Sicilian pond turtle (*Emys trinacris*). For centuries it was mistaken for its mainland cousin, the European pond turtle, until genetic studies in 2005 revealed it as something far older – an island lineage that has followed its own evolutionary course since the end of the last Ice Age. Endemic to Sicily, it survives in scattered ponds, irrigation canals and the few marshes that have escaped reclamation. Its domed shell gleams olive and amber in the sun; it moves through the reeds with the deliberation of something that has already outlived empires.

Like so many of Italy's freshwater relics, its greatest threat comes from us and our tampering: from drained wetlands, polluted streams and pet-shop invaders like the North American red-eared slider (*Trachemys scripta elegans*) that outcompete and sometimes hybridise with native turtles. Yet *Emys trinacris* lives on: a small emblem of Mediterranean patience, basking in the same golden light that has warmed Sicily for millions of years, the last reptilian heir to the vanished inland seas that once covered this land.

Dam Nation

The habitats that sustain Italy's aquatic survivors are among the most damaged in the country. When so much water is funnelled into the few pockets of habitable lowland, control quickly becomes necessity. Centuries of drainage, damming and diversion have left most rivers fragmented and exhausted, their once-lively floodplains reduced to narrow, obedient channels. Pollution, irrigation and hydroelectric schemes have further strangled what were once vibrant, shifting ecosystems. Of all Italy's landscapes, it is the rivers, more than the forests or coasts, that bear the deepest scars of human engineering.

In Italy, trying to orchestrate the waters goes back a particularly long way. The Romans were the first to reshape the peninsula's hydrology on a grand scale: diverting the waters of Lake Velino near Rieti to create the Marmore Falls – still the world's tallest man-made waterfall – taming the Pontine marshes and lacing the fields with aqueducts that carried fresh water for vast distances. At Lake Albano, they cut the world's first emissary tunnel through volcanic rock to regulate its level and irrigate the fields below – a 1.4-kilometre conduit excavated during the Roman siege of Veii in around 396 BCE that still functions today. At Lake Fucino in Abruzzo, once the third-largest lake in Italy, they attempted one of antiquity's most ambitious engineering feats, a 6-kilometre canal designed to drain the entire basin and reclaim its fertile bed for farming. These projects, monumental even by today's standards, revealed a civilisation determined to make water serve its will. Two millennia later, that same impulse endures, though often with less grace and far greater risk. What began as feats of ingenuity has become an industrial reflex: the compulsion to channel, divert and contain.

In Italy alone, of the 1,200 classified rivers, only a handful remain in anything resembling good ecological health. Most are dammed, weired, diverted, drained or polluted by waste, agriculture and industry. The country has more than 500 operational large dams, each over 15 metres high or capable of holding more than a million cubic metres of water. A further 500 large dams stand disused, while more than 90,000 smaller hydraulic barriers – check dams, weirs, obsolete mills, diversion channels – interrupt the country's waterways. At this scale,

Italy's rivers resemble an elaborate industrial plumbing system more than the vital ecosystems they once were. ISPRA (Istituto Superiore per la Protezione e la Ricerca Ambientale) and WWF Italia report that Italy ranks among the most fragmented river systems in Europe, primarily due to the sheer number of small and medium hydraulic barriers, many of them obsolete or unmonitored. Of course, hydroelectric power provides a vital part of the country's clean energy mix, about 15 per cent, but it has often come at huge environmental, and human cost.

High in the mountains straddling Friuli and the Veneto stands one of the tallest defunct dams in the world: Vajont, 262 metres high. In 1963, shortly after it became operational, a colossal landslide from nearby Mount Toc sent 260 million cubic metres of rock and earth crashing into the reservoir. The impact displaced 50 million cubic metres of water, generating a wave 250 metres high that surged over the dam wall. The resulting tsunami obliterated the town of Longarone below, killing more than 2,000 people in seconds. The dam itself remains intact, a grim monument to one of the greatest man-made disasters in history. For those drawn to such places, Vajont is one of the most haunting sites of dark tourism in Europe.

The Tiber – Tevere, the river that gave birth to Rome – remains the longest wholly Italian river, flowing 406 kilometres from the Apennines of Emilia-Romagna to the Tyrrhenian Sea at Ostia. Its watershed once supplied an empire; its name is woven through myth and history. Yet for centuries it has been treated less as a living artery than as an open sewer – walled in, channelled and all but forgotten by the city it created. The embankments that protect Rome from its floods have also severed it from its river.

Still, beyond the reach of concrete walls and hydroelectric dams, long stretches of the upper Tiber and its tributaries continue to flow freely, their banks quietly rewilding in spite of us. And in these wilder reaches, one particular resident has returned – an engineer with rather better instincts than our own, and a talent for healing what we have broken.

Once upon a time, Eurasian beavers (*Castor fiber*) were everywhere in Italy. Their dams shaped the flow of nearly every major river and stream, from the Alpine valleys through the Tiber basin and down to the forests and wetlands of Calabria. For millennia they were the peninsula's true hydraulic engineers, sculpting wetlands, slowing floods, recharging aquifers – until humans decided we could do the job better. By the sixteenth century they were gone, trapped out of existence for their fur, meat and pain-killing secretions, and the country's waterways were left to a single, less competent species of engineer: us. The result was five centuries of straightened channels, drained marshes and rivers that ran faster, lower and emptier than before.

In their absence, an impostor arrived. In the early twentieth century, Italian fur farms imported the coypu (*Myocastor coypus*), a large South American rodent intended to replace the beaver's lost pelt. Many escaped and soon colonised rivers across the country, and without the jaguars to prey on them, their numbers exploded. The coypu filled the beaver's herbivorous niche, grazing reeds and burrowing into banks, but without offering any of the benefits of real engineering. It consumed but did not create, leaving eroded banks and gnawed crops in its wake: a shadow of the ecological master craftsman it replaced.

Then, quite without permission, the beavers came back. Since 2018, they have quietly reappeared along rivers in Lombardy, Friuli, Tuscany, the upper Tiber basin and Molise. No one can agree how. Some point to natural dispersal from Switzerland, Austria or Slovenia; others suspect assistance from well-meaning sympathisers. Whatever the truth, it is one of the most surprising ecological restorations in modern Italian history.

Dutifully, the beavers set to work the moment they arrived. Their dams, now appearing near Tarvisio in the north and on stretches of the Tiber, the Ombrone and Volturno farther south, are once again helping to store water through droughts, trap sediment and create ponds that shelter amphibians, fish, insects and birds alike. As a result, floods spread more gently, reeds and alders return and life rushes back. They are rebuilding in miniature what human policy has spent centuries dismantling. Now, after five hundred years of absence, Italy's best engineers are quietly back on the job.

Their comeback, though, has sharply divided the scientific community. Some view them as uninvited aliens; others as long-lost natives reclaiming their role in the landscape. True to form, Italy has reached the perfect compromise: no one has removed them and the beavers, untroubled by debate, continue to multiply. Their dams and ponds are spreading through forgotten valleys, reviving the ancient conversation between water and life. In a country obsessed with controlling rivers, the beaver is the rare creature that refuses to be controlled. Its return feels less like an invasion than a restoration, the water, at last, reclaiming its memory, one stick at a time.[6]

It's a Sin

There's a river in Abruzzo called the Tirino that is one of the most beautiful sights in all of Italy. Its cold, crystalline waters shimmer between emerald and turquoise, so clear they seem unreal. Beneath the surface, great tufts of water-parsnip (*Berula erecta*) sway like underwater meadows, sheltering native trout, the Apennine yellow-bellied toad (*Bombina pachypus*), and the endangered Italian freshwater crayfish (*Austropotamobius pallipes italicus*), a relic species now extinct in most other rivers. Shoals of Italian riffle dace (*Telestes muticellus*) and horse barbel (*Barbus tyberinus*), both found only in central Italian drainages, flash among the stems. Above them, dragonflies, one of the most efficient predators ever studied, patrol the air – the golden-ringed *Cordulegaster boltonii*, in its Italian form, tracing circles of light over the shallows. Along the willowed banks, grey herons, coots and grebes hunt in silence, as if aware they're moving through one of Europe's last truly living waters.

Years ago, my family and I canoed down the Tirino, astonished by its clarity. Near its source lies Lake Capo d'Acqua, created in the 1960s for irrigation, submerging an old village. On the same trip, I did one of the most unforgettable scuba-dives of my life among the ghostly stone ruins, hovering in near-perfect visibility in a lost submerged world.

Like the Posta Fibreno at the beginning of this chapter, the Tirino is born of the Apennine karst system. Snowmelt seeps into the porous limestone of the Gran Sasso, travelling unseen for years before re-emerging near Capestrano, its purity often hailed as the cleanest in Europe. Yet after only 25 kilometres of bliss, the river ends in tragedy. Downstream lies Bussi sul Tirino, home to one of Italy's worst environmental legacies: the Montedison chemical plant.

For much of the twentieth century, the factory produced weaponised chemicals – mustard gas, phosgene, arsenicals – then dumped its toxic waste into surrounding soil and waterways. In 2007, the site was declared Europe's largest illegal toxic landfill: more than 2 million tonnes of contaminated soil, with poisons still leaching into the Aterno-Pescara basin. The Tirino, one of Europe's cleanest rivers, thus merges with one of its most polluted.

Elsewhere, the story repeats. In Lazio, the Sacco Valley south of Rome has similarly been declared a Site of National Interest (SIN) – a deceptively benign name for zones so poisoned they require national emergency remediation. Once a fertile volcanic plain and cradle of early monastic life, the Sacco was poisoned by decades of pesticide and lindane production. By the time it was banned in the EU, the toxin had already entered the soil, the rivers and the food chain. In 2005, cattle grazing along the river began dying en masse and authorities were forced to intervene, restricting all human activities within the catchment; today, cancer rates here are among the highest in the region, and the area remains among the most contaminated sites in Europe.

Across Italy, around forty such SINs await redemption, from Porto Marghera and Cengio to Augusta Priolo and Taranto. Some have begun recovery; others still languish. As of 2024, only about 13 per cent of contaminated soils and 17 per cent of groundwater have been reclaimed. At this pace, Italy's most polluted landscapes may not be healed until century's end. They occupy less than 1 per cent of the land, yet their effects reach far wider. Most SINs lie along rivers, the sea or, more often, both. Unsurprisingly, it is in these freshwater and coastal habitats that Italy has seen the most dramatic declines in faunal species.[7]

When it comes to long-running environmental scandals, justice in Italy often feels elusive – but perhaps that is finally beginning to change. In June 2025, eleven former executives of the chemical company Miteni were sentenced to a combined 141 years in prison after being found guilty of contaminating more than 100 square kilometres of water and soil in the Veneto region of northern Italy. For years, the company had discharged vast quantities of PFAS, the 'forever chemicals' used in industrial and consumer products, into the local environment, including soils, groundwater and drinking-water supplies, exposing an estimated 350,000 people to substances now linked to cancer, hormonal disruption, infertility and cardiovascular disease.[8]

It is one of the most serious cases of chemical pollution in European history, and the verdict marks the first time corporate executives in Italy have been criminally convicted for such an environmental crime. It is a deeply gratifying result, especially for the valiant local group Mamme No PFAS (Mothers Against PFAS), who spearheaded the campaign, and one that should give hope to the many grassroots campaigners across Italy fighting parallel battles for the right to live in a healthy environment. With combined civil damages exceeding €75 million, and the judgment [in the legal context] also recognising the joint liability of the convicted executives for future cleanup costs, it seems that – maybe, finally – the polluters are being made to pay.

Even so, water endures. It forgets faster than we deserve. It absorbs, dilutes and carries on, reshaping the very world that wounds it. Across the ages, water has been both sculptor and gatekeeper of life. When seas rose and fell (or disappeared completely); when rivers changed course or coastlines fractured, populations split and species diverged. Isolation, repeated countless times, became the rhythm of Italian evolution – its most generative act. Long before the Messinian Salinity Crisis drained the Mediterranean and marooned the ancestors of Italy's freshwater fish, water had already been both destroyer and creator, sculpting life through separation and return.

CHAPTER FOUR

Islands

Oh,
How beautiful is sunset, when the glow
Of Heaven descends upon a land like thee,
Thou Paradise of exiles, Italy!

Percy Bysshe Shelley,
'Julian and Maddalo: A Conversation'

For anyone like me who struggles to grasp the mechanics of evolution, islands are a good place to start. Much of Italy's extraordinary biodiversity can be explained by the simple fact of isolation, and much of its history is an insular one, shaped by islands and the isolations they create. When species find themselves stranded on a finite piece of land surrounded by water, they must adapt quickly or perish. These are the places where evolution runs in fast-forward, sometimes within a few generations. In response to the limits of food, vegetation and predation, whole body types begin to change: large animals shrink, small ones grow, colours shift, limbs lengthen or shorten. If the conditions are right, entirely new and more efficient versions of mammals, birds, reptiles and insects can emerge in a remarkably short time, shaped purely by their surroundings.

Italy's islands still today offer the clearest examples of isolated evolution, but the same forces of adaptation and divergence are visible all around me, and nothing brings this home more than the wall lizards I live among. They're everywhere here, darting across the pavements of Rome, scaling the walls of country houses, vanishing into the rocks of distant islands. No other animal seems to have adapted so completely to every niche of the country, each one resplendent in its own colours and patterned livery. They embody the wildness that runs through Italy's everyday life, a constant, flickering reminder of the wildlife I never grew up with. When I now return to the UK, it's their absence I feel most keenly: the silence between stones, the missed scuffling in the leaf litter.

They are astonishingly quick. You can be looking at one poised on a sunlit rock – still, alert – and the next instant it's gone, vanished into the stone itself. They seem to have an inexhaustible number of escape routes. In moments of boredom, my dogs try hopelessly to catch them in the garden, failing every time. When my son was three, in Parco Sempione in Milan, as we watched them basking on the playground's tuffaceous border stones, he somehow reached out and caught one in his bare hands. Neither of us could believe it, nor have we forgotten it since. The lizard froze, eyes bright and unblinking, before darting free again.

Italy is home to nearly thirty kinds of lizard, from skinks and geckos to the quick, bright-eyed wall lizards that seem to have claimed every stone in the country. These true wall lizards belong to the genus *Podarcis*, a sun-worshipping tribe that first diverged from its relatives some twenty million years ago, when Italy itself was a scattered archipelago rising from the warm, closing waters of the Tethys Ocean.[1] At that time, Sardinia and Corsica formed a great island drifting westward; Elba, Giglio and Montecristo were the crests of submarine ridges; Puglia stood alone as a limestone plateau; and Calabria and Sicily were a ragged cluster at the African edge. In all, perhaps fifty islands and countless islets speckled a shallow inland sea – a geography of separation that would set the pattern for everything to come.

As the islands drifted, collided and reshaped themselves, the ancestors of today's wall lizards spread across volcanic ridges, coral reefs and limestone shores, only to be stranded again as land shifted and seas rose and fell. Cut off from one another, they began to change. Over millions of years this ceaseless cycle of separation and reconnection produced a dazzling range of forms: green, brown, black, striped and, in places, even electric turquoise. What began as a single adaptable lineage has since splintered into an extraordinary diversity: seven recognised species and more than forty subspecies, many confined to a single island or even a single rock.

The most widespread of all, the Italian wall lizard (*Podarcis siculus*), has proved a master of adaptation, colonising almost every region and island of the country. It alone has produced more than fifty described subspecies and local variants, each a subtle response to place from the green-striped forms of the mainland to the cobalt-blue lizards of Capri and the Tremiti. No other reptile in Europe shows such exuberant variety: a living mosaic of evolution still unfolding in plain sight, every population a slightly different answer to heat, colour, vegetation or predator.

Genetic studies have confirmed what naturalists long suspected: these are not mere colour morphs but distinct evolutionary lineages – each a miniature echo of Darwin's finches.* Even now, the story is still unfolding. Recent research has identified a new dark-bellied morph, known as nigriventris, whose appearance may be reshaping the ancient balance of colour forms among Italy's wall lizards.[2] Across Italy's archipelagos, almost every island, even the smallest offshore rock, hosts its own population of *Podarcis*, evolved in seclusion. Rafted, stranded or cut off by rising seas, they have become living proof that even a few kilometres of salt water can redraw the map of life. Seen this way, Italy's islands are micro-laboratories of evolution – small refuges where the ancient rhythm of diversification still plays out beneath the Mediterranean sun.

Nowhere is this more striking than on the Faraglioni sea stacks of Capri, home to one of Italy's most flamboyant creatures: the Capri blue lizard (*Podarcis siculus coeruleus*). Cut off for thousands of years, it has turned a brilliant ultramarine – the very colour once named for coming from 'beyond the sea'. Alpine climbers practising on the Faraglioni were the first to encounter these blue lizards, which, having never met humans before, proved fearless and fiercely territorial, their boldness matching their colour. Naturalists still debate whether the hue serves as camouflage or is simply an evolutionary accident, but it is hard not to see it as the island made flesh: the limestone and light of Capri distilled into reptile form.

Not far to the south, in the Aeolian archipelago, another island-bred lineage clings to existence. The Aeolian wall lizard (*Podarcis raffonei*), described only in 1973 and now one of

---

* Darwin's finches are a group of about fifteen species native to the Galápagos Islands, whose differing beak shapes evolved to exploit distinct food sources. Their adaptive radiation provided one of the first and clearest demonstrations of natural selection in action.

Europe's most critically endangered reptiles, survives on just four tiny refuges in the Aeolian archipelago – a rocky promontory of Vulcano, two small islets near Filicudi and Salina, and Strombolicchio – each holding its own small, genetically distinct remnant. Once widespread across the Aeolians, it has been driven from the larger islands by its more aggressive cousin, the Italian wall lizard. Today fewer than a thousand individuals remain, confined to places so small they can be circled in minutes. On Strombolicchio – the surviving spine of a vanished volcano, rising from the water like a fantasy islet with a solitary lighthouse perched on its summit – these lizards bask on black stone above the sea, a surviving fragment of an island ecosystem now lost from the rest of the archipelago. Fewer than a thousand now survive, scattered across four scraps of rock on the margins of their former range.

Giants and Dwarves

Every summer, Italians flee their overheated towns and cities and migrate en masse to the coast or to one of the country's many islands. Like fish returning to their spawning grounds, there is a ritual element to this annual exodus, an ingrained necessity to reach the beach where a part of the soul has resided since birth. When I married into Italy, I inherited my mother-in-law Elvira's point of origin: Rodi Garganico, a whitewashed limpet clinging to the spur of the boot, jutting into the bright blue Adriatic in northern Puglia. My wife and her sister had spent entire childhood summers in this rough-edged haven, a place at once familiar and distant, but also layered with memories and significance. The name of their grandfather, once the town doctor, lingered over doorways, on street signs and on the lips of sun-stained folk idling in their doorways. Further back, their great-grandfather Felice, an early migrant to the US, had famously returned to Rodi with fistfuls of cash – and with electric light, which he retained the power to switch on and off at will. This proved notoriously useful when certain romantic affairs required crossing the piazza at night, unobserved. In those early years of frequenting the Gargano, it often felt like I was living out a chapter from a novel by Gabriel Garcia Márquez.

The family's annual reappearance in Rodi was celebrated like the return of the long-lost, and even my own arrival as a new face was embraced as if I had always been there. I enjoyed being the village oddity, one of the few foreigners in that part of Puglia (most tourists head further south to Salento), cushioned by the respect of being married into a local family. Perhaps it was the fish-out-of-water spirit of those first summers, but it always felt like stepping into a lost world from another temporal dimension.

The epic drive from Rome only heightened this sense of time travel. Heading east past Tivoli and into the Apennines, one crosses all of Abruzzo, with its dense forests and fantastically precarious mountaintop towns, before reaching Pescara on the Adriatic and turning south. It is one of the great drives of Italy, if not the world. After the brief interlude of coastal Molise, the road bends into the Gargano peninsula, where the scenery changes dramatically: rugged, dry and blanketed in unending olive groves. Finally, the curving descent to the coast, where one or more of the children inevitably risked losing their lunch, before arrival into the warm embrace of Rodi Garganico.

In the 1970s, that uncanny sense of crossing into another world was literally confirmed. Excavations in the *terra rossa* soils of the Gargano peninsula uncovered the fossilised remains of one of the strangest menageries ever found in

Europe – a community of mammals, birds, reptiles and amphibians that seemed at once familiar and oddly distorted. Preserved in the red clays filling ancient karstic caves, these fossils revealed that, around ten million years ago, Gargano was a remote island, cut off long enough for evolution to take some extraordinary turns.

Here, a peculiar fauna evolved in splendid isolation. It included a colossal barn owl (*Tyto gigantea*) and a huge eagle (*Garganoaetus*), both apparently beefed up to prey on the island's comically oversized rodents: giant dormice, fat hamsters, porky pikas and massive shrews that scurried through subtropical forests. In the absence of true carnivorous mammals, birds of prey occupied many of the island's top predator niches, their enormous size reflecting the unusual ecological opportunities created by isolation. Yet danger did not come only from above. On the ground prowled, another of Gargano's apex predators, *Deinogalerix* – the 'terrible moon-rat' – a giant, quill-less hedgehog relative with an outsized head and fearsome teeth.[3]

It seems that the hedgehog ancestor of *Deinogalerix*, like all the creatures that found themselves secluded on Gargano, underwent a series of evolutionary changes that allowed them to exploit new ecological niches of their island sanctuary. For many of these creatures, known collectively as the *Mikrotia* fauna, this meant primarily an increase in size – a well-observed pattern in evolutionary science called island gigantism. Much of the Gargano wildlife would have seemed similar to modern animals we recognise today, only two or three times larger. As if this strange cast of characters weren't enough, completing the rogues' gallery was the gargantuan flightless goose *Garganornis ballmanni*, the enormous otter *Paralutra garganensis*, and the hamster *Mikrotia magna* – an outsized rodent nearly the size of a small dog.

The *Mikrotia* fauna of Gargano likely inflated at first because of a lack of predators, or perhaps a metabolic need for a bigger belly to process a wider range of vegetation. As one species enlarged, so others followed in response. In the case of *Deinogalerix*, it meant becoming a hellishly beady-eyed, hairy, big-mouthed aggressive omnivore and scavenger, unlike almost anything found before or since.

Conversely, the evolutionary drive toward gigantism was balanced by island dwarfism and the need to downsize. Larger animals, marooned as sea levels rose and fell, suddenly had to shrink to more manageable dimensions, relative to the resources available to them. On Gargano, the island's more diminutive but no less curious five-horned, vampire-toothed deer genus, *Hoplitomeryx*, is represented by several species, ranging in size from something akin to a red deer down to the scale of a miniature poodle. Even the remains of a compact crocodile have recently been discovered on what must have been one of the strangest Mediterranean islands.[4]

Minecraft

Gargano was not the only place where Italy's endemic island fauna was evolving at this time. Further west lay the giant island of Tuscania, an ancient landmass formed from Sardinia, Corsica and parts of what is now the Tuscan coast, one of the largest islands to have ever existed in the Mediterranean. Over tens of millions of years, Tuscania was alternately joined to and separated from the mainland, with species evolving and vanishing in rhythm with the shifting seas. About ten million years ago, it too experienced a long period of isolation, during which it developed a unique cast of creatures.

Here, the evolutionary trend went largely the other way. Instead of giants, Tuscania produced dwarves: small antelope-like bovids, known as *Maremmia*; squat buffalo-like creatures called *Tyrrhenotragus*; a primitive giraffid, *Umbriotherium azzarolii*, perhaps reminiscent of an okapi; and the delightfully named *Eumaiochoerus etruscus* – the 'Etruscan pig', a short-snouted, pint-sized boar.

But by far the most famous resident of Tuscania, and the one that has provoked the most debate among palaeontologists, is Europe's best-known fossil primate: *Oreopithecus bambolii*, an unusual Miocene ape long suspected of having affinities with the earliest hominids. When the first reasonably complete skeleton was presented in the 1950s, it caused a sensation, and it has been argued over ever since.

The discovery was thanks to the remarkable tenacity of Johannes Hürzeler, a Swiss palaeontologist from Basel who in 1956 embedded himself in the lignite mines of Baccinello, near Grosseto in southern Tuscany. Living and working side by side with the mining community, he devoted all his energy to uncovering a complete skeleton of this elusive primate, known only until then from scattered fragments across Tuscany and Sardinia.

After two years hard labour in Baccinello, finding only further broken fragments, the mine went into administration and Hürzeler's own contribution from the University of Basel dried up. On 2 August 1958, with his suitcase packed, he was due to leave empty-handed the following morning. Just at the final hour of that final night, however, he was startled awake by the highly excited mine manager telling him that, somewhere 200 metres below them, a tunnel had collapsed and bones were emerging from the vault. Hürzeler rushed down the mine to the fallen tunnel and immediately recognised the unmistakable bones of a complete *Oreopithecus* protruding above him. Luckily, he had the clarity of mind to promptly call a local photographer, while he calmly got to work tracing the skeleton, lest it be damaged in any further collapse of the tunnel. The resulting photos of Hürzeler and the sooty coal miners gazing upward at bones made the covers of newspapers around the world. With incredible care, they were able to safely extract the complete skeleton and this famous fossil was named Sandrone, in honour of one of the miners who so nobly assisted in the acclaimed discovery of what was sensationally declared at the time man's earliest-known relative.[5]

Just like the curious endemic species of Gargano, the key to understanding *Oreopithecus bambolii* lies in how uniquely it adapted to its insular environment. Hürzeler's initial claim that it was the earliest ancestor of man garnered huge attention but was quickly disputed. Without today's fuller knowledge of ape and hominid evolution, his only frame of comparison was with our own lineage. It is now generally accepted that Sandrone's many human-like traits, which Hürzeler correctly identified, were the result of convergent evolution under the pressures of island life. *Oreopithecus* was a peculiarly Tuscanian member of the anthropoid apes that spread across Eurasia at this time.[6] When an ancestor of Sandrone reached Tuscania, it responded to its strange environment by making one of the most human-seeming of adaptations: beginning, at least occasionally, to walk upright. Hürzeler and others were wrong to see him as a direct ancestor of mankind, but we can forgive them. *Oreopithecus* was a clever ape, capable of a convincing, if oddly gaited, Chaplin-esque preview of how we ourselves would one day come to walk.[7]

Sandrone probably thrived for many years on Tuscania with little or no predatory risk, indulging in the peace and freedom to move any way he liked. But his days were numbered. All of these curious Italian Miocene creatures disappeared around five million years ago. Many may even have survived the upheavals of the Messinian Salinity Crisis, when their ocean view became a desert view. When the waters finally gushed back, it was into a land transformed. Rising seas eventually drowned Gargano, while Tuscania was rejoined to the mainland – granting access to the carnivores that made short work of Sandrone, as well as the island's once pint-sized residents.

Island Records

Today, Italy possesses over 50,000 square kilometres of insular territory, scattered across roughly 450 islands and islets, most of them uninhabited. They continue to hide some of Italy's most unique flora and fauna and are now once again providing refuge for the Mediterranean's beleaguered marine life. From the mountainous bulk of Sardinia and Sicily to the smoking cones of the Aeolians, from the limestone outcrops of the Pontine and Tuscan archipelagos to the scattered reefs and stacks along every coast, these fragments form a vast constellation of small worlds fringing the peninsula, each with its own climate, geology and cast of animals.

Of all Italy's islands, Sardinia is both the largest and the most ancient. For tens of millions of years, it has stood apart, drifting westward from the rest of the country like a continent in miniature. Sardinia's story runs counter to the rest of Italy. Once part of the Provençal coast of southern France, it broke away around thirty million years ago and rotated westward, turning slowly on its own axis as the Tyrrhenian Sea opened behind it. Since then, while other parts of Italy have been torn apart, drowned or remade by volcanism, Sardinia has remained steadfast, a survivor of almost every geological convulsion the Mediterranean has known. Today it lies among the most stable pieces of crust in the region, so steady, in fact, that physicists have chosen the disused Sos Enattos mine as a contender to host the Einstein Telescope deep beneath its hills to listen for the faintest tremors in the universe itself.

Sardinia's long isolation has turned it into a museum of living relics and local inventions. Here, evolution did not race forward as it did on smaller islands but settled into patient refinement. The result is an extraordinary fauna, much of it found nowhere else. The Sardinian-Corsican red deer (*Cervus elaphus corsicanus*), smaller and darker than its mainland kin, endures in the oak forests of the Supramonte and along the beaches of Sulcis. The Sardinian wildcat (*Felis lybica sarda*, sometimes *Felis silvestris sarda*) is elusive and solitary, prowling the forested uplands of Gallura; genetic evidence points to introduction from North Africa around the turn of the first millennium BCE, yet it has long since become a distinctive island lineage. Even the Sardinian wild boar retain the look of an older Italy – small, russet, faintly striped – recalling the Maremman boar (*Sus scrofa majori*) that once rooted through oak woods from Tuscany to Lazio; they survive here, at Castel Porziano and, rumour has it, in the darkest corners of the Maremma.

On the high basalt plateaus of the Giara di Gesturi, the Giara pony still runs semi-wild, a stocky, sure-footed horse thought to descend from prehistoric island breeds. The Sardinian long-eared bat (*Plecotus sardus*), described only in 2002 and found nowhere else on Earth, roosts in just three of the island's caves and upland valleys – a quiet, nocturnal emblem of Sardinia's enduring isolation. Even the island's invertebrates have followed their own evolutionary paths. The Nuragic spider (*Amblyocarenum nuragicus*), described only in the 1990s, is a Sardinian endemic as old as the island's stone towers for which it is named. A small, earth-brown trapdoor spider, it lives hidden in narrow burrows under rocks and roots, sealing itself behind a silken lid as if to keep the outside world at bay, a living relic of the island's deep time.

Among the rocks, Bedriaga's rock lizard (*Archaeolacerta bedriagae*) recalls a much older world. Descended from reptiles that colonised the Corso-Sardinian block long before it drifted from Provence, it is stockier and more deliberate than its mainland cousins, haunting granite boulders and wind-carved cliffs. And in the dark of Sardinia's caves lies another hidden world: blind crustaceans, ghostly newts and the delicate cave salamanders – five of Europe's eight cave salamander species occur only in Sardinia – survivors of colder ages whose story we will return to later.[8]

Today, Sardinia remains a world apart – neither African nor European, but something older and uniquely its own: an island that stayed behind while the rest of Italy moved on. Even in modern times its isolation has remained tangible. In the 1960s and 70s, when high-profile kidnappings shook Italy, among them the abduction of Paul Getty's grandson, it was to Sardinia's rugged interior, and to the wildest parts of Calabria, that victims were taken and hidden. These were places where roads ended abruptly, where only shepherds knew every ravine and where the state's reach grew thin. The same geography that sheltered unique bats, cave salamanders and red deer continued, in a darker way, to conceal the kidnapped.

Archipelago of Hope

Between Sardinia and the mainland, a younger generation of islands has taken on a new kind of life. The seven islands of the

Tuscan Archipelago – Gorgona, Capraia, Elba, Pianosa, Montecristo, Giglio and Giannutri – rose from the sea between nine and four million years ago, born of volcanic fires and tectonic uplift. Today, they form Italy's largest marine protected area and one of the most important in the Mediterranean, a sanctuary for the sea's embattled marine biodiversity.

After decades of trawling, poaching and pollution, strict protection, especially around Montecristo, Pianosa and Capraia, has allowed the sea to begin healing itself. Shoals of grouper, bream and barracuda once again patrol the reefs, while meadows of Posidonia seagrass wave in the shallows. Even the critically endangered giant fan mussel (*Pinna nobilis*), the Mediterranean's largest bivalve, once thought doomed by disease and overharvesting, is reappearing on the sandbanks between Elba and Giglio, its long shells filtering the water as they did for millennia. The Mediterranean monk seal, the world's most endangered marine mammal, has even been sighted again among the caves of Capraia, suggesting a tentative return to the islands from which it was once hunted out entirely.

Since 1971, Montecristo has been one of the most strictly protected nature reserves in the Mediterranean, with only a handful of visitors permitted each year, usually for research. Humans have repeatedly failed to colonise it, though not for lack of trying. In the nineteenth century, the Englishman George Watson-Taylor bought the island and built a villa amid its granite terraces, importing goats and exotic plants in an attempt to civilise it; all that remains of his efforts are the feral herds that still wander the slopes. The Italian state acquired the island in 1869, and since then its long isolation has allowed a small, self-contained world to persist almost untouched.

Its granite cliffs and gullies are clothed in lentisk, juniper, rosemary and wild olive. Along the terraces roam the wild goats of Montecristo, descendants of Watson-Taylor's stock, now wholly naturalised. Among the rocks hides the Montecristo asp (*Vipera aspis montecristi*), a pale, elusive island form adapted to the sun-bleached granite and the source of an older legend that Phoenician sailors once hurled vipers at their enemies during sea battles – a tale that earned Montecristo its old name, l'Isola dei Serpenti, the Island of Serpents. Audouin's gulls, pallid swifts and Yelkouan shearwaters nest along the cliffs, and in the clear waters below, fish populations have returned to densities unseen for a century.

Montecristo's notoriety as the setting for Alexandre Dumas's *The Count of Monte Cristo* is now overshadowed by its quieter fame as a refuge for nature – an island so completely protected that it may be the nearest thing left to a living museum of the Mediterranean world before people arrived. Meanwhile, Gorgona, the smallest island, is geologically distinct and humanly curious: still home to Europe's last island penal colony. Around seventy inmates live here with unusual freedom, tending vines first planted by Carthusian monks in the eighteenth century and producing a coveted white wine in partnership with the Frescobaldi family. I sometimes wonder what's the least terrible thing I'd have to do to be sent there.

It's impossible to overstate the importance of the Tuscan Archipelago, the largest protected marine area in the Mediterranean and Italy's only national park composed entirely of islands. In 2021, it became one of the few Mediterranean sites admitted to the IUCN Green List of Protected and Conserved Areas, recognised globally for excellence in management and conservation, one of the rare places where biodiversity is not just surviving, but improving. It is increasingly clear that our seas and oceans are declining faster than almost any ecosystem, largely under the pressure of industrial fishing. But the Tuscan Archipelago serves as the clearest example of how fast aquatic life can return, when given the chance.

The model set by the Tuscan Archipelago National Park could yet be repeated elsewhere: the Aeolian and Egadi islands of Sicily share the same exceptional natural value and would be natural candidates for future national marine parks. Along the Calabrian Arc, the Aeolian Islands are theatres of fire and resilience. Stromboli, the 'lighthouse of the Mediterranean', has been erupting almost without pause for at least two millennia; Vulcano still breathes through its fumaroles.[9] Pumice cliffs hold endemic wildflowers and caper bushes; offshore, dolphins, loggerhead turtles and seasonal runs of tuna and swordfish still pass. To the west, the Egadi Islands shelter some of the Mediterranean's most extensive Posidonia meadows – nurseries for fish and seahorses, anchor for a food web of corals and sponges. Here, *Pinna nobilis* still stands like bronze harps in the sand, and the marine reserve established in 1991 has become a model of recovery, with once-depleted fish stocks now teeming again.

To the south of the Egadi Islands lies the underwater Empedocles volcanic complex. Here in 1831, a massive eruption caused a new island to magically appear overnight. Such was its strategic location that within days, the British, Spanish, French and the Bourbon kingdom of the two Sicilies became engaged in a diplomatic dispute over the island's sovereignty. The British favoured it as a better-positioned alternative to Malta for policing Mediterranean trade, naming it Graham Island after the first lord of the admiralty, Sir James Graham. The Bourbon nobles recognised it as a potentially exciting new beach resort, naming it Ferdinandea after King Ferdinand. Tourists quickly landed to explore the curious new virgin territory with its 63-metre peak and twin lakes. But just as the territorial dispute was heating up, the island deflated and sank back down beneath the waves as quickly as it arose, entirely defusing the tension and saving everyone the trouble of escalating the affair.

Today its summit lies 8 metres below the surface, a colourful seamount teeming with marine life but still outside any formal protection. I once explored it virtually, through a recording by the Marevivo dive team at their centre near Palermo. Ferdinandea's drowned cone is now recognised as part of the Empedocles complex, aptly named for the Sicilian philosopher who first described the world as born of four elemental forces: earth, air, fire and water, and who, legend has it, ended his life by hurling himself into Mount Etna.

Farther south still lies Pantelleria, closer to Tunisia than to Sicily, a volcanic island of obsidian domes and fumaroles, where the wind is so constant that even the trees lean eastward in its grasp. Its interior is a patchwork of caper terraces, lentisk and dwarf olive, while its caves and crater lakes harbour endemic plants, rare bats and migratory birds crossing between continents. The island's landscape, scarred yet enduring, was declared a national park in 2016, the first in Sicily.

And yet the seas around these islands remain largely unprotected. Together, the Aeolians, Egadi and Pantelleria represent some of the last places in the central Mediterranean where natural and cultural heritage coexist in such concentration – each a natural candidate for Italy's next generation of national marine parks. Like the Tuscan Archipelago, they remind us that Italy's future in conservation may depend as much on the sea that surrounds it as on the land itself.

Ice-olation

By the end of the Pliocene, around three million years ago, Italy's familiar outline had begun to emerge. The African plate had long been pressing itself against Europe, raising the Alps and binding the once-separate islands into a single peninsula. At the same time, far across the world, another transformation was reshaping the seas. The Isthmus of Panama closed, severing the Atlantic from the Pacific and disrupting the great currents that had carried warmth around the globe. Ocean circulation faltered; climates cooled.

In the Mediterranean, these shifts brought sweeping evolutionary upheaval. Large baleen whales began migrating north toward Arctic feeding grounds, while the greatest of all sharks, megalodon, unable to follow its prey into colder waters, vanished from the record. Half the world's whale species disappeared, along with a third of its seabirds and sea giants, from dugongs to giant turtles. The warm, tropical seas of the Cenozoic gave way to colder, more turbulent waters, and the stage was set for the most severe cooling in millions of years.

As ice sheets expanded at both poles, sea levels fell, exposing Italy's lowest lands and knitting them together for the first time. Yet this new unity brought a deeper kind of isolation. The Alps to the north and the Apennines through the centre grew white and impassable, sealing the peninsula behind walls of ice. United at last, Italy found itself alone again, an island in all but name, where life would soon be tested by the long winter of the Ice Age. Once more, isolation would become the engine of Italy's evolution.

CHAPTER FIVE

Ice

You see how Soracte remains white
with deep snow, and the struggling trees can
no longer sustain the burden, and the rivers
have been frozen with sharp frost.

Horace, *Odes*

High on a north-facing slope of the Madonie Mountains in northern Sicily stands one of the world's rarest trees. The Sicilian fir (*Abies nebrodensis*) looks like a misplaced Alpine emigrant, a conical Christmas tree marooned among scrub and limestone under the southern Mediterranean sun. Smaller and darker than its northern cousins, with short, curved needles and a crown thinned by drought and wind, it stands out as though it had wandered here by mistake. Indeed, it is in fact the last survivor of a lineage that once spread across the Mediterranean highlands during the cold Pleistocene, having diverged from its silver fir (*Abes alba*) relative more than a million years ago. It endures in a small cluster of limestone ravines, rooted among great rockfalls where the air remains just cool enough and the soil too thin for beech or oak to encroach.

Its cones yield so few fertile seeds that most of its offspring survive only through the careful hands of botanists. Like a lost tribe, barely thirty mature trees remain in the wild. Fenced off and watched beside a hiking trail, they are lonely relics of an age when both temperatures and seas lay far lower than today, and when Sicily still looked north toward a frozen continent.

At the time of the last glacial maximum, around twenty thousand years ago, Italy was almost unrecognisable. The Alps were a single, seamless expanse of ice, their valleys filled to the brim and their frozen veins spilling far out toward the Po Valley. From Turin to Verona, the land ended in walls of blue-white ice that calved meltwater torrents into a tundra of gravel and sedge. The great northern lakes – Maggiore, Como, Garda, Iseo – lay buried beneath glaciers kilometres thick, their basins gouged deeper by repeated glacial advances.

To the south, the Apennines glittered with smaller glaciers that clung to their highest peaks – Gran Sasso, Maiella, Velino – feeding crystalline rivers that wound through wind-scoured steppe. The lowlands were treeless and open, swept by cold winds from the north. Forests had withdrawn into a few sheltered ravines and coastal enclaves, leaving much of the peninsula carpeted in tundra grasses and herbs: artemisia, saxifrages, dwarf willow, juniper and patches of hardy mountain pine. Herds of reindeer, wild horse, aurochs (the wild ancestor of domestic cattle) and steppe bison grazed the plains, while chamois and ibex picked their way across the exposed limestone ridges. In the far north, mammoths still lingered beyond the Alps, but their southern cousins had long since vanished. The Adriatic was not a sea but a vast, grassy plain, a corridor of Ice Age prairie stretching from Venice to Puglia, where cranes wheeled over thawing marshes and wolves and Stone Age humans hunted the herds that grazed there.

Even further south, the Mediterranean had withdrawn far from its present shores. Sardinia and Corsica formed a single mountainous island, and the Tyrrhenian coast lay as a line of cliffs and dunes. Only a few corners of Italy – the mild Tyrrhenian slopes, the coastal pockets of Campania and Calabria – remained unfrozen. In these scattered enclaves of green, life waited out the cold: birds, insects, small mammals, reptiles and amphibians – the ancestors of many of the species that would one day recolonise Europe when the ice began to retreat. Others went further still, evolving to the chill and scarcity of their southern refuges, becoming the uniquely Italian endemics that still endure there today.

When the thaw began, the conifer forests that had once blanketed the Mediterranean withdrew northward, following the retreating cold, but the Sicilian fir could not follow. Adapted to the thin chill of glacial winds, it found itself marooned on its mountain refuge as warmth gradually rose from the coasts below. Its companions – silver firs, black pines and highland spruces – gradually vanished from Sicily as the climate warmed, leaving this one specialised fir alone in a climate no longer suited to it.

Here, on its exposed ridge, the Sicilian fir survives more by luck and human care than by nature's design: fenced from goats, sheltered from fire, coaxed to reproduce in nurseries. It is one among many Italian endemics now classed as critically endangered – stranded survivors scattered across the peninsula, living remnants of a world once far colder than today. They persist where they first evolved in isolation, beyond the easy reach of cultivation and development, in the last pockets of Italy still shaped more by climate than by control.

Igloo

The Ice Age was the defining period of Italy's natural history, the crucible in which its extraordinary biodiversity was forged. When most of Europe lay buried beneath ice and tundra – the treeless, frozen plains of the north – the Italian peninsula remained a rare sanctuary: a long, temperate bridge between the frozen north and the deserts of Africa. While glaciers sealed the Alps and spread across the continent, Italy, along with its southern neighbours Spain, Greece and the Balkan peninsula, became one of Europe's great glacial *rifugia* – a haven of warmth and life. As the tundra spread across northern Europe, animals faced a stark choice: adapt, move or die.

But the 2.6 million years of the Pleistocene were far from a continuous icehouse. Over that time, the ice advanced and retreated in rhythmic pulses, more than twenty in all, at least eleven of them major, freezing and thawing the landscape each time. During the great glacial advances, the Alps vanished beneath kilometres of ice and the Po Valley froze into tundra. In the warmer interludes, with temperatures often higher than those today, forests of oak and beech reclaimed the valleys, rivers filled with meltwater and elephants, deer and early humans. Throughout these almighty climatic swings, Italy's geography proved its greatest gift: a long, mountainous peninsula projecting into warmer seas, blessed with a cavernous underworld full of hiding places, and spared from the worst of the ice. Glaciers and permafrost crept over its Alpine summits and the highest

Apennine ridges, but the south and the islands remained largely ice-free.

It was an age defined by vast waves of migration and extinction, shaping the modern distribution of countless species, including ourselves. When sea levels dropped, new land bridges emerged, most dramatically across the Adriatic basin, which transformed from sea into a vast grassy plain. When the glaciers finally began to retreat in earnest around fourteen thousand years ago, the human survivors in Italy were joined by newcomers from further north and east – an essential infusion of new genes remembered today as the Villabruna turnover, named after a cave in the Veneto where several Late Glacial skeletons have been analysed. Genetic studies show that these individuals carried a distinct ancestral signature, with links stretching into the Near East, that spread rapidly across the continent and became the dominant genetic foundation of later European populations. Remarkably, it remains a major component of modern European ancestry.[1]

As humans were reshaped by these renewed connections, so too was the wider ecosystem. The same climatic shifts that opened new routes for people also opened corridors for animals. These shifts created easier migration routes into Italy, funnelling waves of creatures from Africa and Asia into the peninsula. Alongside the bears, wolves, deer and horses familiar to us today came the great Pleistocene megafauna: elephants, rhinos, hippos, cave bears, giant hyenas, sabre-toothed cats and cave lions. At various points during the Ice Age, all of these species found refuge in Italy, where they adapted and evolved, while further north they were replaced by cold-adapted giants such as woolly mammoths and musk ox. With each cycle of glacial advance and retreat, animals shifted in rhythm with the climate, moving south when the ice closed in, then expanding northward again as the world warmed.

Trunky

Although Italy's lands were by now largely united as a single peninsula, the climatic rhythms of the Ice Age continued to enforce the old laws of island evolution. Each glacial pulse redrew the map of the country. When the seas fell, islands became bridges; when they rose, the bridges were cut and evolution resumed its insular experiments. Among the fantastic beasts of the Pleistocene – sabre-toothed cats, giant hyenas, cave lions, woolly rhinos and bison, all of which at some point were present in Italy's folds – few traced these changes more faithfully than the proboscideans.

Nowhere in Europe has yielded a richer or more continuous record of elephants and mammoths than Italy. From the Valdarno in Tuscany and the volcanic plains of Lazio to the lignite mines of Umbria and the karstic caves of Sicily, their bones chart two million years of migration, adaptation and extinction, a fossil archive of giants shaped by the climate. Over that immense span, multiple elephant species appeared and vanished across the peninsula, evolving into some of the most extraordinary endemic forms the world has ever seen.

One of the earliest to arrive in Italy was the southern mammoth (*Mammuthus meridionalis*), a fur-less, warmth-loving giant whose first fossil was discovered in the Valdarno region of Tuscany in 1825. In 1954, a near-complete adult male skeleton was unearthed at Madonna della Strada in Abruzzo, today mounted in the Forte Spagnolo in L'Aquila, where its titanic proportions can be fully appreciated:

nearly 4 metres high, 6 metres long, and weighing an estimated 11 tonnes – about twice the size of a modern Asian elephant. Another extraordinary site is Pietrafitta in Umbria, where the remains of eight southern mammoths were uncovered in a lignite mine alongside hippos, rhinos, bison, deer and dozens of other species. Together, these finds provide some of the best snapshots of Pleistocene Europe, showing the richness of Italy's Ice Age fauna about 1.5 million years ago.

As temperatures dropped further in the mid-Pleistocene, around a million years ago, the steppe mammoth (*Mammuthus trogontherii*) entered the scene. Adapted to arid, open grasslands, the steppe mammoth gradually replaced its southern ancestor. From this lineage, the iconic woolly mammoth (*Mammuthus primigenius*) eventually evolved, capable of colonising even the coldest corners of Ice Age Eurasia.

The steppe mammoth was followed into Italy during a subsequent interglacial by the equally colossal straight-tusked elephant (*Palaeoloxodon antiquus*), whose remains have been found throughout central and southern Italy, especially in Lazio. Straight-tusked elephants were ubiquitous across the peninsula from around eight hundred thousand years ago until as recently as forty thousand years ago, and Italian specimens are sometimes distinguished as the ancient Italian elephant (*Palaeoloxodon antiquus italicus*).

Both the steppe mammoth and the straight-tusked elephant seem to have reached an understanding over their living arrangements in Italy, coexisting across the peninsula for tens of thousands of years. They explored nearly every corner of the land, and when sea levels fell, even pushed out to the islands. During the early Ice Age, much of Calabria south of Catanzaro was still submerged, with only the

Aspromonte, Serre and Capo Vaticano rising as islands. Across the water, Sicily, too, was fragmented, with only the northern and central uplands, plus the ancient Hyblean volcanic islands near Syracuse, standing above the waves. These Calabrian stepping stones gave adventurous elephants and other animals a route to Sicily. As sea levels continued to drop, the fragments of Calabria and Sicily merged, eventually linking up not only with each other but, during the lowest sea stands, with Malta as well.

Seeking ever warmer lands, the straight-tusked elephant followed its trunk as far as it could, eventually finding itself marooned on Sicily and Malta when the land bridges disappeared beneath the waves. With no apparent predators, limited space and vastly different vegetation, one of the world's largest elephants responded to its new island home by shrinking into the world's smallest: the Sicilian dwarf elephant (*Palaeoloxodon falconeri*).

Over many generations, this extraordinary transformation saw each new elephant born smaller than the last, dwindling from more than 4 metres at the shoulder to just a single metre – about the size of a large dog. That would be the equivalent of a fully grown human shrinking to the size of a rhesus macaque. Its tusks, once long and straight, gradually curled upward and outward, a feature some palaeontologists argue links it more closely to the steppe mammoth. The best evidence comes from the Spinagallo Cave near Syracuse, which yielded thousands of bones and several near-complete skeletons, proudly displayed today in Milan, Catania, Syracuse and even in casts as far afield as the Smithsonian in the US. It is a textbook case of insular dwarfism and a contender for the cutest creature to ever walk the earth, if not for the suspicion that the discovery of its skull in classical times inspired the Cyclops legend, its trunk socket mistaken for a fearsome third eye.

Dwarf elephants are perhaps the most emblematic of the Pleistocene fauna that inhabited the Mediterranean islands during the Ice Age. Their remains have also been found on Malta, Crete and Cyprus, where each population underwent its own evolutionary race to the bottom. When Sardinia was periodically connected to, or at least swimmable from, Tuscany, either the steppe or southern mammoth also made the crossing and duly shrank to island size. As on Gargano or Tuscania millions of years earlier, mammals, birds and reptiles that reached these islands either ballooned to enormous proportions or dwindled into miniature forms.

On Sicily, the dwarf elephant shared space at different times with a similarly reduced dwarf hippopotamus, a pint-sized fallow deer, the largest dormouse ever to scurry the earth, a giant tortoise and a massive swan that loomed over nearly all of them. Later still, a midsized dwarf elephant, *Palaeoloxodon mnaidriensis*, joined this Ice Age menagerie, the product of a separate evolutionary downsizing, perhaps in response to larger threats and increased competition on the island.

It's remarkable to think that our hominid ancestors once lived alongside many of these magnificent beasts of the Pleistocene. But their coexistence was often fraught, and for the proboscideans in particular, it proved catastrophic. Few creatures have fared worse when forced to share space with humans. In Italy, the last of the great elephants vanished around forty thousand years ago, as the last glacial advance intensified and modern humans spread across the peninsula, marking the end of one of the longest continuous elephant lineages in Europe. They would not return again until 280

BCE, when Pyrrhus of Epirus – he of the famous Pyrrhic victory – brought twenty of them to Lucania, modern-day Basilicata. The Romans, astonished at these towering animals, nicknamed them *boves Lucanos* – Lucanian cows. Hannibal's audacious march over the Alps with thirty-seven elephants sixty years later would ensure their lasting place in legend, bringing the Republic close to collapse.

Of all the Ice Age animals, I'd probably most like to see a dwarf elephant, but scientists are aiming higher. Since the 1980s, efforts have been underway to resurrect the long-extinct mammoth through genetic engineering, using the living Asian elephant as a biological stand-in. The goal is to create a 'Pleistocene Park' of Ice-Age megafauna, supposedly to restore lost ecosystems and slow global warming. Yet mammoths went extinct through a combination of climate change and human pressure – both of which still exist. None of the stated motives for bringing them back truly hold up.

I would love nothing more than to see a woolly mammoth or rhinoceros alive again, but this feels like science for science's sake: impressive, expensive and ultimately a little vainglorious – akin to colonising Mars while ignoring Earth. It is undeniably thrilling, more ambitious and more attractive to investors, yet it risks distracting from the less glamorous and far more urgent task of conserving what we already have. The harder, humbler work is to keep what remains alive. In Italy's case, though the elephants are long gone, there remains a great deal still to save.

Sleep like a Dormouse

In the wake of the final melt, beginning about nineteen thousand years ago, came both renewal and loss. Some species expanded; others vanished forever. As Lucretius observed, nothing is born without something else passing away. Yet Italy remained, as it had been throughout the long seesaw of the Pleistocene, a bridge between worlds – the peninsula where life endured the cold and from which it returned when the ice released its grip. Hedgehogs, hares, chamois, lizards and salamanders, among many others, owe their survival to Italy's fortunate geography, its mountains, coasts and islands offering warmth when the rest of Europe froze. From these refuges, life spread outward again, repopulating a thawed continent.

Some species, however, had become so finely adapted to their Italian strongholds that they never left. Today, Italy hosts the highest number of endemic vascular plants (about 1,300) and animals (around 4,800) in all of Europe, many of them living relics of the Ice Age. Some trace their ancestry back to lineages that once ranged widely across ancient Eurasia, now reduced to small, isolated pockets where they have evolved independently. Their greatest concentrations lie in Calabria, Sicily and Sardinia, thanks to their prolonged isolation, but examples can be found in nearly every region. With so many hidden corners, the discovery of new endemics in Italy remains a regular occurrence even today: in 2025, botanists described *Festuca silana*, a new grass species found only on the Sila plateau of Calabria.

Calabria and its rugged interior play host to a variety of species that have remained isolated for an exceptionally long time. Hemmed in by the Ionian and Tyrrhenian seas, this wild landscape has acted as one of Italy's most important refuges, allowing creatures to evolve in relative seclusion and fostering an extraordinary level of endemism, including many of the country's small mammals.

Dormice, despite their name, are not mice at all but members of a far older and more distinguished lineage, the *Gliridae*, an ancient family of Old World rodents more closely related to squirrels than to rats. They first appeared in Europe around fifty million years ago, about the same time whales were taking their first tentative steps into the sea. For much of the ensuing epochs they were among the most diverse rodents of ancient Europe, reaching their peak diversity during the Miocene. Highly adaptable, dormice spread across Europe, Asia and Africa, and around thirty species still survive today. As nocturnal forest climbers with unfussy diets and an admirable fondness for sleep, they perfected the art of hibernation, dozing for half the year or longer. The English phrase 'to sleep like a baby' is affectionately and, far more accurately in my opinion, rendered in Italian as *dormire come un ghiro*: 'to sleep like a dormouse'.

The dormouse, in particular, has been enjoying sweet dreams in Italy for a very long time. Members of its family are among the peninsula's earliest mammalian settlers, with fossil relatives present here for almost forty million years. Their remarkable adaptability served them well across Italy's shifting landscapes. On the ancient Miocene island of Gargano, at least five endemic species evolved, including the formidable *Stertomys laticrestatus*, literally the 'giant snoring mouse'. This heavyweight probably rivalled Sicily's Ice Age mega-dormouse, *Leithia melitensis*, which reached the size of a domestic cat. Another insular form arose in Sardinia, where the darker, larger *Eliomys quercinus sardus* evolved in isolation – an island form of the garden dormouse now considered extremely rare and possibly extinct.

Dormice also adapted to the cold. One of the surviving lineages is the bushy-tailed forest dormouse (*Dryomys nitedula*), a woolly-coated species that favours high altitudes and dense woodlands. Today, it remains one of just five surviving European dormouse species. Its range stretches across Eastern Europe, the Balkans, the Caucasus and deep into Russia and Central Asia, with the Italian and Swiss Alps marking its westernmost outpost.

However, there is a particularly curious relict population that persists far to the south in the forests of Pollino, Sila and Aspromonte in Calabria: *Dryomys aspromontis*. Isolated for so long that it has diverged from other forest dormice by about a million years, it is now recognised as a distinct endemic species. Like a thief in the night, it wears a dark mask across its eyes and bounds through the tree canopy at altitudes often exceeding a thousand metres above sea level. Representing an ancient lineage, it has weathered every climatic mood swing of the past million years, quietly thriving in Calabria's secluded uplands, well out of reach of modern life.

While the Calabrian forest dormouse keeps to the high, quiet woods, its larger cousin has not been so lucky. The European edible dormouse (*Glis glis*) got its name for a reason – for millennia it has been considered a delicacy. The Romans regarded it to be among the most delicious of foods, either as an hors d'oeuvre or sweetened with honey for dessert. Dedicated glirarium vessels were described by Rome's great agricultural writer Varro, in which the animals were kept and fattened for special occasions. Galen and Pliny the Elder referenced them in their writings, and in Petronius' *Satyricon*, at the infamous feast of Trimalchio, they appear 'sprinkled with honey and poppy seeds' (*glires melle ac papavere sparsos*). Recipes also feature in the world's oldest cookbook, Apicius' *De re Coquinaria*. Even after the fall of Rome, the culture of eating them lingered on in Italy, with recipes quietly passed down through the centuries and still discoverable today in more obscure corners of the internet.

During a recent anti-mafia operation near Reggio Calabria, police discovered 235 dormice in the freezer of a suspected henchman, with others alive in cages being fattened up. Meals of dormice, it turns out, remain central to the traditional rites of the powerful 'Ndrangheta mafia. Usually pan-fried or stewed, they are consumed on special occasions: initiation ceremonies, discussions of a hit, the sealing of new alliances or the negotiation of disputes. Being a protected species, a dish of dormouse symbolises both a disregard for national law and the primacy of local custom. Forget the Hollywood image of corpulent mobsters around heaped plates of spaghetti and meatballs, the real ones eat small, furry rodents. The illegal night-time hunting of dormice is so widespread in parts of Calabria that in just one district near Catanzaro, as many as twenty thousand are believed to be captured annually. Probably best not to complain too much if you are ever served one though or you may find yourself sleeping like a dormouse forever.

The Calabrian forest dormouse is one of several ancient mammals that survived the Ice Age in Italy's diverse landscapes and still thrive today. Sharing part of the same southern territory is the Calabrian black squirrel (*Sciurus meridionalis*), which evolved from the familiar Eurasian red squirrel but, like the forest dormouse, has been isolated in Calabria's forests for so long that it is now recognised as a distinct endemic species. With its striking black-and-white coat and imposing size, this southern squirrel has thrived, growing around 30 per cent larger and heavier than its northern cousins.

Once confined to the high forests of Pollino, Sila and Aspromonte National Parks, its range is now expanding northward, perhaps following the spread of its favourite tree, the Calabrian black pine. Recent sightings have even placed it at sea level in Basilicata's coastal pine forests. Meanwhile, the Eurasian red squirrel (*Sciurus vulgaris*) is steadily pushing southward. What will happen when these long-separated ancient cousins finally cross paths remains to be seen. But if it comes down to a fight, my money's on the southern heavyweight in black.

Like dormice, squirrels have deep evolutionary roots; but in Italy their future now depends less on ancient history than on the pressures of the present. The invasive grey squirrel (*Sciurus carolinensis*) – the same species that decimated red squirrel populations in the UK – was introduced to Piedmont in the early twentieth century, from where it spread into Liguria and Umbria. Recent interventions, including non-lethal chemical sterilisation, have slowed its advance across the Apennines, and for now it remains contained in the north. More worrying is the unchecked spread of another invader, Finlayson's squirrel (*Callosciurus finlaysonii*), released through the pet trade in Basilicata in the 1980s and now advancing into both Campania and Calabria.

Hares almost certainly recolonised much of Europe from Ice Age refuges in Italy and the Balkans and one of their most distinctive survivors is the endemic Italian hare (*Lepus corsicanus*). With its smart white underfur, it remains prevalent in Sicily and persists in scattered populations across southern Italy, reaching as far north as western Tuscany. Once widespread across the entire peninsula, its mainland presence has become increasingly fragmented due to habitat loss, hunting and competition from the larger European hare (*Lepus europaeus*), introduced in modern times.

Unlike its more adaptable cousin, the Italian hare enjoys protected status as a legally untouchable endemic rarity – though, being nearly indistinguishable from each other at a distance, this protection is almost impossible to enforce. Curiously, it also disdains interbreeding with outsiders: unlike many other closely related species, it flatly refuses to hybridise with the European hare, holding fast to its own lineage. Dignified though this may be, it has made the Italian hare all the more vulnerable to decline and possible extinction.

In Cold Blood

The glacial refuges of the peninsula during the Pleistocene were especially critical for cold-blooded amphibians, perhaps the most vulnerable creatures of all in that era. These are among Europe's oldest lineages, many of them stretching back long before the age of the dinosaurs. Remarkably, nearly half of the amphibian species found in Italy are endemic relict survivors, their cousins having vanished everywhere else on the continent. It seems this land preserved more of the damp, shaded, humid habitats these animals so desperately require.

My original muse and inspiration, the spectacled salamander, is one of the finest examples of these Ice Age holdouts. Alongside Lanza's Alpine salamander (*Salamandra lanzai*), the Italian crested newt (*Triturus carnifex*) and the Apennine yellow-bellied toad (*Bombina pachypus*) and others, it endured the climatic convulsions of the past two million years by clinging to cool, forested refuges. None of these venerable creatures are built for speed or long-distance migration. Instead, they stayed put, finding in Italy the exact niches they needed – and never leaving.

In a handful of clear, high-altitude pools in Calabria's shaded uplands lives one of Italy's most surprising survivors: the Calabrian Alpine newt (*Ichthyosaura alpestris inexpectata*), whose Latin name rather wonderfully translates as 'the unexpected alpine fish-lizard'. Endemic to the southern Apennines, it too descends from a population stranded here as the glaciers retreated, marooned in the cool plateaus of the Sila and Aspromonte. Its blue-marbled males and orange-bellied females still glide through moss-rimmed pools where the water stays cold all year. It is, quite literally, an alpine 'fish-lizard' where no alpine creature ought logically to be – an amphibian spirit of the Ice Age adrift in the Mediterranean south.

By far the most intriguing of Italy's relicts are the eight species of cave salamanders (*Speleomantes*), nearly all of which are found exclusively in Italy, five of them only in Sardinia. These shy, elusive creatures are the last European representatives of a once-diverse group of lungless salamanders that today numbers about five hundred species in the Americas, with a single outlying species discovered in Korea in the early 2000s. Genetic studies suggest they diverged from their parent group around the end of the dinosaurs. Living a predominantly subterranean life, they have found stable refuges in the karstic caves and rocky crevices of the northern and central Apennines, the Maritime Alps and Sardinia, with small outposts in San Marino, Monaco and Provence.

Similar to geckos, cave salamanders are equipped with adhesive toe pads that let them scale sheer rock and cling to the slick walls of their subterranean realm. Though diminutive in size, they are adorned with striking colours and iridescent markings. Seen up close, their skin shimmers as though dusted with gold leaf, kept glistening by the permanent patina of moisture that varnishes their supple little bodies.

At night, they slip from their caves to hunt in the dripping forests outside, snapping up insects with lightning-fast strikes from their remarkably long, sticky tongues. By day they retreat deep into the cool dark underworld, conserving the precious moisture that keeps them alive.

Unlike most amphibians, *Speleomantes* possess no lungs at all. Instead, they breathe entirely through their skin, oxygen diffusing straight into their blood while carbon dioxide seeps out the other way. This adaptation allows them to thrive in the humid, oxygen-rich crevices of caves and fissures, squeezing into the tiniest cracks of the underground world. But it comes at a cost: if their delicate skin dries, they asphyxiate. A sheen of moisture is all that stands between them and suffocation.

This sensitivity to humidity and temperature shifts means even slight changes to their habitat can threaten the survival of cave salamanders. In recent years, however, a far greater danger has emerged. The pathogen *Batrachochytrium salamandrivorans* (Bsal), a fungus that thrives in damp environments, attacks salamanders by consuming their skin. The infection causes sores, lethargy, loss of appetite and ultimately death. A related fungus, *Batrachochytrium dendrobatidis* (Bd), has already devastated amphibians worldwide, driving more than two hundred species to extinction, including Costa Rica's once abundant golden toad (*Incilius periglenes*). These fungi spread with ease: their spores linger in water bodies for long periods, remaining both contagious and virulent. In recent years, entire populations of fire salamanders in the Low Countries and Germany have been wiped out, and scientists dread the news of its further spread.

Please don't search for cave salamanders. In Italy, the disease threatens some of Europe's most ancient and irreplaceable survivors. For the rarest species, like the Sette Fratelli cave salamander (*Speleomantes sarrabusensis*), found only in a single location, extinction would come swiftly. The best way we can protect them is by leaving them undisturbed and by not risking contamination of their fragile refuges. It should be enough simply to know that these miraculous creatures endure, hidden in their damp caves and forested crannies, as they have for millions of years, and, if fortune allows, will continue to do for many more.

Staying Alive

Lake Pilato, high in the Monti Sibillini National Park on the border between Le Marche and Umbria, is a place steeped in myth, mystery and tragedy. According to local legend, it was here in this remote mountain wilderness that Pontius Pilate wandered in exile after being recalled from Judea in disgrace by the Emperor Caligula. Tormented by guilt over his role in Christ's trial and crucifixion – and perhaps also by the ignominy of being shamed by the most shameless of emperors – Pilate is said to have hurled himself into the lake's icy waters, seeking final respite from the horrors of his past.

Today, this fragile glacial tarn hosts an even more endangered creature: the Marchesoni fairy shrimp (*Chirocephalus marchesonii)*, a species completely unique to Italy, its entire population bound to the seasonal freeze–thaw rhythms of a single high-altitude basin in the Sibillini Mountains. Its cousin, *Chirocephalus sibyllae*, survives just a few kilometres away in another mountain lake, together making these mountains an unlikely world capital of endangered fairy shrimps. Both belong to an ancient Eurasian lineage of ephemeral-pool crustaceans, whose drought-resistant eggs can be carried vast distances on the feet or feathers of migratory birds – a mechanism thought to explain how their ancestors reached the Apennines after the last Ice Age. Both species are also critically threatened by shifting hydrological cycles and warmer summers that shrink their already precarious habitats, sounding a clear alarm about the changing times.

Fossils of fairy shrimp (*Anostraca*) trace their lineage back to the early Palaeozoic, more than 350 million years ago, and their basic form has changed remarkably little since. Rather than competing in predator-rich waters, they became specialists of shallow, temporary pools – fleeting habitats where fish and other large hunters cannot survive. Among the planet's most resilient creatures, fairy shrimp have survived every major extinction since multicellular life began. Today, they persist on all seven continents, from the frozen wastes of Antarctica to the scorching deserts of the Sahara. That evolutionary pedigree has armed them with formidable survival skills, allowing them to outmanoeuvre death at every turn.

At Lake Pilato, their life is as precarious as ever. From November until May, the basin lies buried beneath ice and snow. When the thaw arrives, meltwater refills the twin pools of the mountain, which from above appear like a pair of blue sunglasses set into the grey limestone. This seasonal flood triggers the hatching of the fairy shrimps' 'eggs' – more precisely, drought-resistant embryos known as cysts, locked in suspended animation within the sediment. Within days, they transform into vivid coral-red adults, barely a centimetre long, skimming upside down through the icy water.

Their time is fleeting. In just a few months, which in normal years is between June and September, they must feed, grow, reproduce and deposit the next generation of cysts, before the lake dries or freezes once again.

Sadly, these are no longer normal years. Before 2017, Lake Pilato had only dried out a handful of times in recorded memory. Since then, dwindling winter snowfall and rising temperatures have caused the basin to desiccate completely every summer, with water levels dropping alarmingly low even in May and June, when meltwater should be at its peak. Geologists also suspect that the earthquakes of 2016 further fractured the mountain's structure, allowing what little meltwater there is to drain away more quickly. Fortunately, the fairy shrimp has long since learned not to put all its eggs in one basket: its cysts can endure years of drought, lying in wait until the lake returns, a strategy that has helped it survive for half a billion years.

Hedging its bets, only a fraction of *Chirocephalus'* cysts ever hatch, always leaving a hidden reserve buried in the sediment should the current generation fail to complete its cycle. No one knows exactly how long they can remain in hibernation, but other species of fairy shrimp have reappeared after decades of absence, springing suddenly back to life when long-dry pools refilled. The greater risk now lies not only in drought, but in the boots of summer hikers, trampling across the desiccated lake bed and crushing the dormant cysts lying in wait for better times.

During the calamitously dry summer of 2024, park authorities fenced off the basin, but the outlook for Marchesoni's fairy shrimp remains precarious. For a creature that has weathered half a billion years of planetary upheavals, it would be a cruel irony if its downfall came not from climate catastrophe alone, but from the careless footsteps of passing hikers. Fortunately, 2025 brought much more rain and meltwater, a fortuitously good year for the fairy shrimp in which it was able to complete its full reproductive cycle, but the future remains very uncertain.

The extinction of a tiny, high-altitude crustacean might not seem significant to most people. Out of sight and mind, Marchesoni's fairy shrimp lacks the charisma that fuels conservation campaigns, and even if one were mounted, it would struggle to capture attention. But when an ancient survivor as resilient as the fairy shrimp disappears, alarm bells should be ringing. Not just for the loss of a unique species, but for what it signals about our world. The drastic shift in Lake Pilato's hydrological cycle in just a few short years is a warning writ large: the landscapes we take for granted are changing fast. If the cold, high places that once stored and released water can no longer do so, then downstream, where we rely on it for our lives, the flow will only dwindle.

Ice Remembers

It is important to remember that for most of its history, Earth has been a warm, ice-free world. Out of 4.5 billion years, perhaps only 15 per cent has been spent in what scientists call an icehouse state – periods when one or both poles were locked beneath permanent ice. The rest of time belonged to the greenhouse: tropical seas lapping at polar shores, coastlines drowned beneath warm, swollen oceans. That we live in an age with frozen caps at both ends of the planet is, geologically speaking, an anomaly, and it is within this rare cold chapter that almost the entire human story has unfolded.

The past twelve thousand years, the whole of human civilisation, from the first fields to the present moment, have taken place within one of the brief interludes of warmth that punctuate an Ice Age. This respite is known as the Holocene. Under natural conditions, the next great glaciation would likely begin roughly ten thousand years from now. But that seems improbable today. For the first time, it is not Earth's orbit or the tilt of its axis that will decide the planet's next chapter, but us.

The extraordinary flora and fauna Italy inherited from the Ice Age, thanks to the country's unique geography and long role as Europe's refuge, is now being forced to adapt at an almost impossibly fast pace. Life has done this many times before, but never with such speed. In the past, shifts in climate unfolded over millennia, giving species time to migrate, evolve or find sanctuary. Today the rate of change outstrips their ability to respond. Some may climb higher or move north, but for many, especially Italy's rare endemics, there may soon be nowhere left to go. The legacy of the Ice Age, a story written over tens of thousands of years, risks being erased in mere decades.

The same forces are reshaping the land itself. The Alps, Italy's glittering northern bastion, still rise a few millimetres each year as tectonic uplift continues and the crust rebounds from the loss of its ancient ice. Since the last glacial maximum, tens of trillions of tonnes of ice have melted from the range; freed from that weight, the mountains are slowly lifting, floating upward like corks released from water.

Yet while the peaks rise, their lifeblood is draining away. Snowfall across the Alps has fallen by more than a third in the past century; glaciers have lost a fifth of their surface since the 1980s; and one great Dolomite glacier, Marmolada, now thins by several centimetres a day. Italy's snow-water reserves plunged by over 50 per cent in a single winter. Each summer, the rivers that once thundered from the Alps and Apennines now falter weeks earlier, their snow and glacier melt reduced to trickles. The springs that feed Italy's lakes, farms and aqueducts are weakening – the same dwindling flow that starves Lake Pilato and its fairy shrimp. In 2022, the Po reached its lowest recorded flow in more than two centuries, a stark warning that the glacial-fed arteries of Italy's water system are running dangerously thin. What was once the steady heartbeat of the mountains, the seasonal pulse of meltwater, is becoming erratic, fading toward silence.[2]

The miraculous discovery of the spectacled salamander in Pereta, Tuscany, 4 April 2020.

'Ciro' (*Scipionyx samniticus*), Italy's first dinosaur fossil and one of the most detailed in the world. *Photo by Leonarda Vitola*

The Carrara marble quarries, Tuscany. *Photo courtesy of Alessandro Colle / Shutterstock*

The Alpine ibex (*Capra ibex*), almost entirely descended from the last surviving population in the Gran Paradiso National Park. *Photo courtesy of Gianluca Damiani*

One of the world's rarest bears: the Marsican brown bear (*Ursus arctos marsicanus*). *Photo courtesy of Gianluca Damiani*

The Italian wolf (*Canis lupus italicus*), one of Europe's greatest conservation recoveries. *Photo courtesy of Gianluca Damiani*

The Apennine chamois (*Rupicapra pyrenaica ornata*), often considered the most beautiful of all chamois. *Photo courtesy of Gianluca Damiani*

Etna: a splendid lava fountain during the eruption of 13 November 2023.
Photo by Vincenzo Miconi

Hot springs, Bagni San Filippo, Tuscany.

The Devil's Footprints at Roccamonfina volcano in the Roman Magmatic Province.

The gladiator crab, Rome. *Photo courtesy of Emanuele Biggi*

The European owl moth, a miraculous relict of the Ice Age, found only around Monte Vulture. *Photo courtesy of Renato Spicciarelli*

The Apennine yellow-bellied toad. *Photo courtesy of Federico Crovetto & Shutterstock*

The Etruscan excavated roads of Pitigliano, Sorano and Sovana cut directly out of the volcanic tuff (*tufo*). *Photo courtesy of Travelvolo/Shutterstock*

The falls of Marmore in Umbria.

The town of Pitigliano, built from tuff. *Photo courtesy of SF William / Wikimedia*

The slightly unreal Tirino River, Abruzzo. *Photo courtesy of Luigi Filice*

The blue lizard of Capri. *Shutterstock*

Deinogalerix koenigswaldi, the hellish hedgehog from Miocene Gargano. *Courtesy of Roman Yevseyev / Other Worlds*

The Appian Way, paved with basalt, flanked by the dying pines of Rome. *Photo courtesy of Lorenza Campanella / Wikimedia*

The critically endangered Supramonte cave salamander. Please do not attempt to look for it! *Photo courtesy of Franco Andreone*

The Calabrian forest dormouse. *Photo courtesy of E. Fava, Associazione Ge.Co*

The world's loneliest fairy shrimp, *Chirocephalus marchesonii. Photo courtesy of Marco Scolastici and Parco Nazionale dei Monti Sibillini*

Lake Pilato, high in the Sibillini Mountains, home to *Chirocephalus marchesonii. Photo courtesy of Alessandro Rossetti*

The hermitage of Santo Spirito a Maiella, Abruzzo. *Photo courtesy of Gianly87 / Wikimedia*

Spot the hermitage, San Bartolomeo in Legio, Abruzzo.

The lakes of Monticchio with the Abbey of San Michele Arcangelo housing the Museo di Storia Naturale del Vulture, Basilicata.

CHAPTER SIX

Wood

Naturam expellas furca, tamen usque recurret.
You can drive Nature out with a pitchfork, but she will keep coming back.

Horace, *Epistles*

Of all the wonderful creatures I encounter around our farm in the Maremma, none delights me quite as much as the crested porcupine (*Hystrix cristata*), the *istrice* (EE-stree-che, as Italians say it). Long before I ever saw one, I knew they were here – their long black-and-white quills turned up everywhere: along forest paths, under olive trees, caught in the fence by the vegetable garden. At heart, they're fairly ordinary rodents, but their extravagant armour turns them into something else entirely. The few times I've caught a group of them running ahead of my headlights on the road, quills fully raised, they made me think of punks in wedding dresses hurrying to church, bristling with attitude and improbable elegance. When startled, they rattle their quills in defiance, a rustling protest from a creature that has survived both glaciations and empires.

We have at least two families living in deep burrows along the slopes that fall to an overgrown ravine, hemmed in by cork and holm oaks, trees whose roots twist through the same soil as the porcupines' tunnel. After an early learning period of having rigid spikes painfully removed from their flanks, my dogs have long since learned not to waste too much energy chasing or berating them, however much they might want to.

Italy is the only country in Europe where these marvellous spiky animals still live. Fossils show that porcupines once roamed the peninsula during the late Ice Age, appear to have disappeared during the coldest phases of the last glaciation, and returned when the climate warmed, along with the forests that now shelter them. Whether they came back by their own wanderings from North Africa, or more likely with a little help from the Romans, who carried them, painfully, across the sea, they have stayed for good.

Until recently, the crested porcupine was Italy's largest terrestrial rodent, before it was surpassed by the beaver, which began returning in around 2018 after an absence of more than five centuries. Today, *Hystrix cristata* thrives from Sicily to the Alps, and although it is now second in size, it is undoubtedly the best dressed of the continent's rodent world. Monogamous by nature, the species mates for life and shares the work of digging elaborate burrows, where both parents raise their young, who in turn often return to the family home, reproducing there for generations to come. Feeding on roots, bulbs and fallen fruit, they live as quiet foragers and useful engineers: their tunnels are borrowed by foxes and badgers, while their quills adorn magpies' nests – and my own bookshelves. When we came to establish the Tuscany Environment Foundation a few years ago, it was this combination of resilience, fidelity and quiet ecological usefulness that led us, unanimously, to choose the crested porcupine as our symbol.

Reintroduced by humans into a human-made world, the crested porcupine is a fitting symbol of the Holocene itself: a survivor of prehistory thriving in an age of design and dominance, proof that control has never been absolute, and that life, when given half a chance, finds its way back. And as the climate warmed and the ice released its grip, that resilience played out on a continental scale. The world that returned was not one of glaciers and grasslands, but of leaves – a slow, unfolding reclamation by trees.

The Rebound

Italy began recovering from the Ice Age sooner than most of Europe, its recolonisation by plants and animals already well underway by fifteen thousand years ago. In those first temperate millennia after the ice withdrew, the land was reborn. Beech and oak advanced up the valleys; pines spread along the coasts; birch and hazel filled the clearings where the glaciers had retreated. Rivers ran freer, soils deepened and Italy turned green again – an expansive mosaic of forest stretching from the Alps to Sicily, save for the highest peaks and the most sodden plains.

By the time of the Holocene, the last twelve thousand years, most of the great Ice Age megafauna – mammoths, woolly rhinoceroses and giant deer – had vanished, while the last cave lions, cave bears and hyenas faded away in the shadow of retreating glaciers. Afterward, a different cohort endured: aurochs, ibex, chamois, red deer, wild boar, brown bear and wolf – the survivors of the deep freeze, species that adapted rather than vanished. Together they formed a bridge between the vanished Pleistocene world and the warmer, human-shaped landscapes of the Holocene, a lineage that, apart from the aurochs, persists in Italy today.

As the ice withdrew and the forests returned, humans began to take charge of the land, felling, burning, draining, sowing and building, yet even in the most altered landscapes, traces of the wild persisted beneath the trees. The forest became both refuge and frontier, a living margin between wilderness and settlement. From the chamois on the highest limestone peaks to the wolf haunting the reclaimed plains, from the porcupine's burrows to the sparrow's eaves, life found shelter among the woods and learned to live alongside the works of its human controllers.

Not Seeing Wood for Trees

In their day, the Romans harboured a famously ambivalent attitude toward the forests that still cloaked much of Italy. They revered them as places of divine presence, yet feared them as dark, lawless realms beyond the reach of order. Among the most formidable was the Selva Ciminia, a vast woodland of oak, beech and hornbeam that spread across the volcanic Cimini Hills, dividing Rome from Etruria. Livy called it *horrida atque invia* – a dreadful and almost impassable wilderness. In 310 BCE, when the consul Quintus Fabius Rullianus proposed crossing it to outflank the Etruscans, the Senate, unnerved by its reputation, forbade the attempt. Forests represented both mystery and menace; centuries later, the massacre of three Roman legions in the Teutoburg Forest confirmed that the great woods were the empire's most treacherous frontier.

But Rome could not live without them. Timber was the lifeblood of the empire: homes, ships, scaffolding, siege engines – all consumed it. Hardwoods such as oak and ash supplied beams, ploughs and spear shafts; beech furnished furniture and charcoal; larch built aqueducts and bridges; and the resinous Calabrian pine was prized for the pitch that waterproofed Rome's fleets. No region escaped the axe. Tuscany's hills were stripped to supply warships during the Punic Wars; Calabria's pines were felled for beams and planking. Each trireme, Rome's warship powered by three tiers of oarsmen, devoured huge quantities of timber, and the arsenals of Ravenna and Misenum (the Roman naval base at Cape Miseno, Bay of Naples) burned through entire forests. By the fourth century CE, Rome's bathhouses alone required mountains of firewood, and when local supplies dwindled, the city began importing timber from North Africa, the Balkans and Anatolia.

These clearances decimated the habitats of countless species that had once roamed freely across the peninsula. Even in Roman times, one senses a quiet awareness of what was being lost. Virgil, writing in the *Georgics*, evokes a world still rich with life: monk seals along the coast, wild swans near Mantua, lynx, aurochs, cranes, deer and wild goats. Latin authors frequently mention boar, roe deer and ibex, once so plentiful they filled the hunting parks and private reserves of aristocrats. But as forests fell and fields spread, this great fauna began to fade.

The *venationes*, arena hunts staged for public entertainment, only accelerated the decline. Though famed for importing lions and leopards from Africa, the Roman games were just as destructive to Italy's own wildlife. Emperor Gordian I, in a single spectacle, reportedly offered up a thousand bears, two hundred deer, one hundred wild goats and dozens of boar and ibex – many of them likely captured from the Apennines. Even provincial towns joined in the carnage: at Minturno, near today's Lazio–Campania border, forty bears and sixteen deer were slaughtered in a single festival in 249 CE.[1] No landscape, however remote, was beyond Rome's appetite for blood sport.

Birdlife may have fared somewhat better, especially in the wetlands and lagoons of places like Ravenna, where pelicans and storks were once common sights. Yet others were less fortunate. The northern bald ibis (*Geronticus eremita*), described by Pliny the Elder and known to nest on Alpine cliffs, was relentlessly hunted, its eggs a delicacy on princely tables. By the Renaissance, even memory of the bird had faded, and not until the nineteenth century was its identity rediscovered on the shores of the Red Sea. Today, a small flock once again traces that ancient route, guided by microlight from Austria to the wetlands of Orbetello in southern Tuscany, an ancient migration, reborn.

Forest Food

Still, the forests endured because they fed both body and spirit. Sacred groves – the *luci* – were scattered across the landscape, dedicated to deities such as Diana, Persephone and Bacchus, and watched over by the priestly Arval Brotherhood. To cut their trees was to invite divine wrath. Forests also sustained life in humbler ways. Acorns fattened Rome's pigs in oak pastures known as *silvae glandariae* and, in lean times, fed people too. The Twelve Tables, the earliest written code of Roman law, even included *de glande legenda,* 'on the gathering of acorns', a reminder of how closely food security was rooted in trees. Even today, Italy's most famous cured hams, from the Prosciutto di Parma to the culatelli of the Po Valley, trace their lineage to those ancient acorn forests of Emilia-Romagna, when most of the pork for Roman tables was raised in the oak woods around Modena.

From the oak pastures came other gifts that would long outlast Rome itself. The hazelnut, *Corylus avellana*, a native of Italy's own temperate woods, takes its name from the Campanian town of Avella, east of Naples on the road into the mountains, whose groves were famed as the finest in the empire. Long gathered from the wild, it was among the first forest fruits to be deliberately cultivated in Italy, thriving in the cool, fertile soils of the Apennine foothills. From these southern beginnings it spread northward, a quiet domestication that would, centuries later, feed an industry satisfying our hunger for hazelnut spread, far removed from its ancient roots.

The walnut – *Juglans regia*, from *Jovis glans*, meaning Jupiter's nut – flourished in the deep alluvial soils of Emilia-Romagna and Campania, its timber carved into furniture and its nuts scattered at weddings as emblems of fertility and fortune. Further south, the almond, brought by Greek and Phoenician traders, flowered across Sicily and Puglia, the first tree to wake after winter and the first to feed. Together these nut trees – hazel, walnut, almond – became Italy's earliest woodland orchards: crops rooted in the forest margins rather than the sun-baked groves of olive and vine, bridging the world of wild foraging and cultivation.

From these ancient woodlands arose one species destined to transform Italy's uplands: the chestnut. Native to the forests of southern Europe and western Asia, with ancient refuges stretching from the Caucasus to the Apennines, it has been present in Italy since prehistoric times and was systematically cultivated from the Bronze Age onward. It was the Romans, however, who spread it widely across their empire, recognising its worth as both timber and sustenance. In their wake, whole mountainsides were remade in its image: vast *castagneti* – ancient chestnut groves managed for generations – spreading below the beech line, especially on the volcanic slopes of Amiata and Roccamonfina, and through the humid Tuscan Apennine valleys of Lunigiana, Garfagnana and Mugello. In these uplands where grains faltered, the chestnut became known as the bread tree – the source of flour and daily sustenance for both shepherds and peasants, staving off hunger for millions well into the twentieth century. Its sweet, starchy nuts could be milled into flour or roasted; its wood made strong beams and barrels; its bark yielded tannins for leather; and its coppices ensured a steady supply of firewood.

On the eastern slopes of Mount Etna stands its most famous monument, the Castagno dei Cento Cavalli, the 'Chestnut of a Hundred Horses'. Said to have sheltered the Queen of Aragon and her retinue from a storm, it is thought to be up to four thousand years old. By the eighteenth century, its girth was reputed to exceed 60 metres, securing its place among the great trees of the world.

Still today, the chestnut harvest marks one of Italy's most cherished rites of autumn. Forget the city street vendors hawking inferior imports; head instead to the hill villages of Arcidosso or Santa Fiora on Monte Amiata, to Pontremoli in the Lunigiana, or to the cobbled lanes of Marradi and Castiglione di Garfagnana, where families still roast and grind the new crop in the glow of wood fires. In those places, the *metati* – the old stone drying huts for making flour – still exhale the scent of sweet smoke, keeping alive a bond between people and tree that reaches back to Rome itself. Among all

Italy's trees, none speaks more clearly of the long entanglement between forest and civilisation.

What Did the Romans Ever Do for Us?

When undertaking the herculean land reclamations needed for expansion, the real problem for the Romans was the law of unforeseen consequences. As they dug canals, dredged rivers and channelled water to the sea, they also cleared surrounding forests to fuel their fires and construct their civilisation.

The very acts meant to tame the landscape often undermined it. Without the forests, they lost the stabilisers of the land – the natural sponges that absorbed rainfall, reduced erosion and regulated river flow. Floods became more frequent, even in the dry season, washing soil and debris back into the very channels the Romans were so desperate to keep clear. Wetlands refilled, rivers silted up faster and farmland was damaged by sudden inundations or left parched in drought. Rome found itself in a feedback loop of ever-larger interventions, just to hold on to earlier gains.

Only now are we beginning to grasp the true magnitude of what forests do. Long treated as mere sources of fuel and timber, they are in fact vast engines of stability – breathing systems that cool the air, summon rain and bind the soil. Through processes sometimes described as the 'the biotic pump', forests draw moisture inland from the sea, creating the very clouds that feed their own rivers. Their canopies slow the rain, their roots hold it, their soil releases it in measured flow. Each woodland, however small, is a local climate machine – a sponge, a filter, a regulator of rivers and weather alike. When trees fall, the water runs wild; when they return, the cycle steadies again.

We are just starting to see how deeply the fate of water is tied to that of the woods, something the Romans, for all their brilliance, discovered too late. Their drained marshes and stripped hills brought floods, silted rivers and the slow exhaustion of the land. We stand once again at that same threshold of realisation: to lose the forest is to lose the climate itself.

But despite all their bewildering feats of engineering and ruthlessly efficient administration, the Romans never fully subdued the Italian landscape. By the third century CE, the golden age was already waning. The centuries-long climate optimum that had rendered the Mediterranean basin so fortuitously fertile began to unravel, giving way to droughts, instability and climatic stress. Over time, the land itself pushed back. Centuries of relentless cultivation, mining and logging had stripped its strength. Soils grew thin and vast stretches of once-productive countryside slid into exhaustion. Roman writers themselves sensed the decline: Columella, Rome's most influential writer on agriculture, warned of fields worn out by neglect, while Pliny the Elder lamented that the land was being 'plundered rather than cultivated'.[2]

In the centuries that followed Rome's dominion, one species after another slipped quietly from the Italian forests. By then, the great aurochs had already disappeared, having once thundered through the oak groves of the Po Valley. Wild horses soon vanished from the peninsula, their bones turning up in the ashes of Etruscan hearths to mark their passing. Bears and lynx lingered longer, retreating into ever-smaller refuges until both vanished from much of the peninsula. Only the Marsican brown bear endured, confined to a single pocket of the Apennines, while the lynx disappeared entirely for a time, surviving only beyond Italy's borders. The beaver endured a little longer, trapped for its pelt and scent, before disappearing. Yet now, in our own time, all three have returned: bears once more roam the Alps, lynx slip silently back across the frontier, and beavers gnaw again at the willow banks of the rivers. Despite all the upheavals of the past two thousand years, several of the forest's ghosts have found their way home again.

The aurochs itself, ancestor of all domestic cattle, vanished in 1627, when the last cow died in a Polish forest. In a curious modern turn, conservationists are now attempting to back-breed its likeness by selectively crossing primitive cattle. Among the key breeds are Italy's long-horned Maremmana and the rugged southern Podolica, living heirs to the wild giant that once roamed these same plains.

The forests, above all, are the clearest terrestrial symbol of how Italy's landscape has changed over the past twelve thousand years – and how, even when forcefully removed, they always find their way back. They have risen and fallen like tides, advancing and retreating in rhythm with human presence. In the beginning, they covered almost everything: a vast, breathing canopy stretching from the Alps to Sicily. As people multiplied, they retreated and fell beneath the axe and the plough. The Romans then accelerated their destruction, and over the next two millennia the woods swelled and shrank in step with war, abandonment and recovery, until today, when they have returned to cloak roughly 40 per cent of the country.

Though far from the endless forests of Scandinavia or Slovenia, this still marks one of the greatest woodland recoveries in Europe. Italy's forests, however, are unlike any other: a living mosaic compressed into a single, slender peninsula. From the spruce and larch of the Alps to the beech and silver fir of the Apennines, from Mediterranean oak and cork groves to the carob, myrtle and wild olive of the southern coasts,

Italy contains every major forest type of Europe within its borders. And those ranges of species often coexist within a single region, even in the far south of Calabria. In the space of a few hundred kilometres, one can walk from glacial conifer woods to subtropical evergreen thickets, crossing more climatic zones and biogeographical regions than in any other European country.

Rare Canopies

This extraordinary overlap has made Italy the continent's botanical crossroads and one of its greatest arboreal sanctuaries. Within Europe, few countries host such a variety of native trees: at least 140 species, more than almost anywhere west of the Balkans. Some, like the Sicilian fir, Lobel's maple and the Etna birch, are totally unique. The structure of Italy's forests is equally diverse – dense high-altitude canopies, open oak pastures, cork scrub and coastal maquis, each shaped by centuries of disturbance and renewal. This is the legacy of a land that served first as Europe's refuge during the Ice Ages and later as its great corridor of return, where northern, eastern and African floras met and mingled in perpetual exchange.

In the Apennines, where the beech still reigns supreme, a few sacred pockets endure much as they did twelve thousand years ago, when the first forests reclaimed the land after the glaciers withdrew. These *faggete vetuste*, the primeval beech woods, are among the oldest surviving forests in Europe, the last true witnesses of a bygone age. On the border between Tuscany and Emilia-Romagna lies the Foreste Casentinesi, a vast wooded heartland and national park that, along with the Tuscan Archipelago, is among the few Italian sites recently added to the IUCN Green List – rare places where biodiversity and ecological integrity are measurably improving. Within it, the Sasso Fratino reserve shelters 500-year-old beech trees not far from the Franciscan and Camaldolese monasteries of La Verna and Camaldoli, whose monks have lived amid these woods for more than eight centuries. Declared Italy's first strict forest reserve in 1959, and closed to all but scientific study, Sasso Fratino has remained entirely untouched by human hands ever since.

Together with twelve other sites stretching from Abruzzo to Aspromonte, Sasso Fratino forms part of the Faggete Vetuste d'Italia – Italy's contribution to UNESCO's network of Ancient and Primeval Beech Forests of Europe. Though smaller than the vast tracts that still survive in the Carpathians or Białowieża, Italy's stands are the most southerly and ecologically diverse of all, preserving the genetic heartland from which beech recolonised the continent after the Ice Age. They mark both the cradle and the frontier of Europe's forest history.

Further south, in the deep folds of Val Cervara within the National Park of Abruzzo, Lazio and Molise, dendrochronologists have counted more than five centuries in a single trunk – trees that sprouted before Columbus ever set sail, silent witnesses to the centuries when bears and wolves still ruled the hills. In Lazio, the ancient beeches of Mount Cimino, once part of a vast forest that awed the Romans, rise from volcanic tuff, their roots drawing sustenance from the same ash that built Rome. And in Puglia's Gargano Peninsula, my mother-in-law's native ground, beeches grow improbably close to the sea, among the lowest and most maritime beech forests in Europe.

In Calabria, two forests tell twin stories of endurance. On the high plateau of the Sila stand the Giganti della Sila:

fifty-eight monumental Calabrian black pines (*Pinus nigra* var. *calabrica*), an endemic subspecies that once blanketed the uplands between Cosenza and Catanzaro. Some soar 40 metres high and have stood for nearly four centuries, withstanding lightning, wind and waves of human clearance. They are all that remains of a vast pine forest felled in the 1940s, when Italy, broken by defeat, paid part of its war reparations to the US in timber. Countless trunks were shipped across the Atlantic to New York, where they served as scaffolding for the city's new skyscrapers – a living forest transformed into the skeleton of another.

The giants of the Sila survived only because the local landowner, Baron Mollo, refused to let them be cut. Today, they are protected within both the Sila National Park and the Fondo per l'Ambiente Italiano's (National Trust for Italy) Riserva Naturale Biogenetica dei Giganti della Sila, a rare case of overlapping national and civic guardianship. They rise like a natural cathedral: colossal trunks straight as columns, bark dark and fissured, crowns converging into a fan vault canopy far above the ground. When I visited, I had my first sight of Calabria's endemic black squirrels scratching and hopping around their titanic trunks and branches.

In the limestone fastness of the Pollino stands something rarer yet: the Bosnian pine (*Pinus heldreichii*), a relic of the glacial world and one of the oldest verified living trees in Europe. One ancient patriarch, nicknamed Italus, sprouted more than twelve centuries ago, when Charlemagne was still a boy and Lombard kings ruled the plains below. Its bleached bark and twisted limbs gleam ghost-white against the cliffs, like driftwood from another epoch. Between the Calabrian black pines of the Sila and the Bosnian pines of the Pollino, Calabria shelters both the youngest and the oldest forests of Italy, one reborn from the wreckage of war, the other untouched since the last ice began to melt.

The Last Forest of Rome

Closer to the coast, much of Italy's once-mighty evergreen forest has vanished, cleared for fields, roads and resorts. Only a few fragments survive intact. The finest lies on the Roman littoral, within the presidential estate of Castel Porziano, Italy's answer to Windsor Castle, which shelters an arboreal and faunal pantheon unlike any other on the Mediterranean mainland. According to legend, this was the shore where Aeneas first landed from Troy, the mythic seed of Rome's foundation. Protected for more than seventy years, it preserves one of the last great remnants of Mediterranean coastal forest, a living echo of the green fringe that once bordered the entire Tyrrhenian shore. Declared a state nature reserve in 1999, Castel Porziano harbours extraordinary biodiversity: over a thousand plant species and nearly three thousand species of large animals and birds, including perhaps among the last populations of the native wild Maremman boar.

Here the great evergreens of antiquity still hold their ground: vast holm oaks engulf the ruins of Roman villas; immense oleasters, the wild ancestors of the olive, stand gnarled and silvered with age; and nearer the dunes, Phoenician juniper, myrtle and lentisk cling to the sand, sculpted by salt and wind. But it is the pine, the Roman stone or umbrella pine (*Pinus pinea*), that defines both this coast and the city beyond. Admired since antiquity for its timber and nuts, the umbrella pine became a civic emblem only in the late nineteenth and early twentieth centuries, planted in regimented rows along boulevards, piazzas and the Appian

Way to lend Rome its distinctive skyline. Its resinous scent and spreading crowns became as integral to the city's identity as domes or ruins.

Yet the very symbol of Rome is dying. Across Lazio and the Tyrrhenian coast, tens of thousands of umbrella pines are succumbing each year to a microscopic invader: the pine tortoise scale (*Toumeyella parvicornis*). Feeding on the trees' sap, it excretes a sugary honeydew that fosters black mould, choking photosynthesis and leaving the pines vulnerable to beetles and drought. Native to the southeastern United States, the insect arrived in Naples around 2014, most likely through the ornamental plant trade, and has since advanced relentlessly northward. In Castelporziano alone, more than 100,000 pines have perished in just five years. Walking there today feels like wandering through a vast arboreal cemetery: whole stands blackened and skeletal, the air heavy with resin and decay.

The die-off has provoked anger and disbelief across Italy. By 2019, with nearly 80 per cent of Rome's pines affected, emergency treatments became compulsory, though they can only slow the pest's advance. Many of the city's trees were already nearing the end of their natural lifespan; their loss will permanently alter the skyline that inspired poets, painters and Ottorino Respighi's *Pini di Roma*, his orchestral portrait of Rome's stone pines. As the singer Antonello Venditti once put it, '*Come i pini di Roma, la vita non li spezza*' – 'Like the pines of Rome, life does not break them'. Yet if not life, then sadly a small insect is now doing just that.[3]

At Castel Porziano, however, the crisis has sparked renewal. The estate's ecologists see in the devastation an opening to restore the forest's original complexity. By 2026, over 300,000 saplings from 54 native species will have been planted – not as a replacement monoculture but as a rewilding, a slow return to the mixed oak, olive and myrtle woodlands that once ruled the Roman coast. As in Yellowstone, wolves have recently re-established themselves within the reserve, re-imposing the 'ecology of fear' that keeps deer and boar in motion and allows saplings to rise again. Castel Porziano's pines may not survive the century, but their passing will not be in vain: from their fallen giants, a truer forest may yet be born – the last green lung of Rome, and perhaps the most hopeful.

Yet Castel Porziano is only the most visible example of a far wider crisis. Across Italy, trees are under siege. The pine tortoise scale is just one of many invasive species overwhelming the country's forests, most of which, unlike Castel Porziano's, lack protection or resources. Nearly all of Italy's coastal palms have been wiped out by the Asiatic palm weevil; Tuscany's cypresses are hollowed by fungal canker; and chestnuts still suffer the lingering effects of blight. In the Dolomites, vast stands of Norway spruce have fallen victim to bark beetles supercharged by rising temperatures and storm damage. When Storm Vaia struck in 2018, its 200 kilometres per hour winds felled more than 14 million trees and the deadwood became a breeding ground for beetle outbreaks that turned whole mountainsides red – a living scar of climate change in the north, as invasive species ravage the south.

Seen together, Castel Porziano and the Dolomites tell the same story: forests collapsing under pressure, yet capable of regeneration. What is happening here is, in truth, happening everywhere. Seen like this, all forests are symbols of renewal. True old-growth forest is now exceedingly scarce, and this scattering of ancient woods – the beeches of the north, the pines of the south, the oleasters, holm oaks and cork groves of the coasts – together represent the smallest fraction of

Italy's forests. Nearly all of Italy's forests have at some point been felled, burned, replanted and left to grow again; yet through all the centuries of cutting and clearing, they have never surrendered completely. Even now, new forests rise where fields are abandoned, spreading once more over the wounds of history, with Italian forest cover rebounding by as much as 100,000 hectares each year.

And as the trees return, so too does the life that depends on them. From the mountains to the plains, the old rhythms are stirring again: deer browsing at the edges of new woods, wild boar rooting through the leaf litter and wolves slipping back into the margins of their ancestral range.

A Real Boar

Emblematic of this rebound is one of Italy's rarest mountain survivors, the Apennine chamois (*Rupicapra pyrenaica ornata*), a slender goat-antelope of the high mountains, often hailed as the most beautiful of all its kind. When the glaciers withdrew and the forests rose again, it followed the cold upward, abandoning the valleys for the cliffs and high meadows above the tree line. Though it now grazes in the open heights, its life still depends on the forests below, their roots anchoring the slopes, their shade feeding the springs that rise to its pastures.

Nearly lost a century ago, when only a few dozen animals remained in the Abruzzo, the chamois has rebounded thanks to protection and the quiet return of woodland. From that last refuge in the Abruzzo, Lazio and Molise National Park, it has been reintroduced to the Maiella, Gran Sasso and Pollino massifs. Today, more than three thousand individuals roam once again along the spine of the Apennines, a living bridge between the wooded heart of the mountains and the bare summits above.

A quieter emblem of the same renewal is the Italian roe deer (*Capreolus capreolus italicus*), whose fortunes have long mirrored those of the forest itself. Once common in the lowlands, it retreated to the last inviolate pockets of woodland: the presidential estate of Castelporziano near Rome, the Umbra Forest of Gargano in Puglia, and the uplands of the Pollino in Calabria. A few small populations have since been reintroduced in the Tolfa Hills, Cilento, Aspromonte and southern Tuscany. Less than a thousand remain, making it one of Italy's most endangered large mammals. It is an animal that vanishes when the trees fall and returns when they rise again, its fate bound, more than any others, to the presence of the forest. Hybridisation with northern roe deer, introduced through restocking programmes, now poses its gravest threat.

Far from the mountains where most of Italy's deer now roam, descendants of reintroduced Central European stock, a tiny remnant native lineage endures in the coastal woodlands of the northern Adriatic. The Mesola, or Italian, red deer (*Cervus elaphus italicus*) once ranged widely across the lowland forests and dunes between the Po and the sea. Cut off and nearly lost, it now survives only in the Bosco della Mesola, a last pocket of ancient oak and holm oak at the delta of the Po, and since 2023, in a new refuge in the Serre of Calabria. Long thought to be a northern strain, genetic studies have revealed it to be something older and more southern, a close cousin of the Sardinian-Corsican and even North African deer. Smaller and darker than its continental kin, it stands as the final glimmer of Italy's vanished lowland wilderness.

On Sardinia, a close island cousin, the Sardinian-Corsican red deer (*Cervus elaphus corsicanus*), tells a parallel story of survival and rediscovery. Separated from the mainland since the last glaciation, it evolved into a compact, russet form, adapted to the maquis and evergreen oak forests of the islands. By the 1970s, it had almost vanished, driven to the brink by hunting and deforestation. Thanks to decades of protection and reintroduction, its bugling now echoes once more through the cork oak woods of Sulcis, Monte Arcosu and the valleys of southern Corsica – a fragile but heartening recovery, and a reminder that, even here on the islands, the ancient pulse of the Holocene forest can still be heard.

Sadly, when I went to try to track down the Sardinian-Corsican red deer myself, among the pristine dunes of Piscinas on the Costa Verde, that pulse was hard to hear. The dunes, among the largest in Europe, rise 60 metres high and stretch for kilometres inland – a landscape so unspoiled it feels outside of time. But between the golden ridges, I found veins of red water cutting through the sand: toxic run-off from the long-abandoned lead and zinc mines of Montevecchio and Ingurtosu. Following the Rio Irvi upstream, its rust-coloured flow led me to the shuttered shafts where groundwater, flooding the tunnels since the 1990s, has dissolved oxidised metal salts into an acidic brew. The island's oldest rocks, more than five hundred million years old, now seep poisoned water, slowly bleeding from wounds that refuse to heal. Despite years of promised cleanup funds, there has still been no attempt to contain the flow. I saw no deer that day, only the scar of an island still paying the price for its buried wealth.[4]

The wild boar's story of recovery is one of excess and interference. Italy once had several local forms of wild boar (*Sus scrofa)*, each adapted to its landscape: the small, wiry *cinghiale maremmano* of central Italy; the *apenninico* of the interior; and the island boars of Sardinia and Sicily. Most of these native strains have been turned over by the introduction of larger northern and eastern European boar during the twentieth century, released by hunters seeking to 'improve' the species. The hybrid offspring spread with astonishing success, replacing the lighter native types and multiplying far beyond their historical numbers. Yet traces of the original lineage persist in the wilder corners of Sardinia, within the protected walls of Castel Porziano and in a few isolated forests of the Maremma. The *cinghiale* – wild boar – remains a symbol of Italy's wilderness, but it is a wilderness refashioned by human hands: an ancient animal remade for the modern age.

Around my farm in the Maremma, both roe deer and wild boar are constant presences, forest residents of the surrounding Mediterranean forests and maquis interacting continuously – the true tenants of the land. At dawn, the deer step delicately from the oaks, ears flicking, vanishing at the first sound of movement. Their tracks trace the same paths each morning through the olive terraces, which then become my own pathway for walking and foraging. By night, the boar take over: whole families rooting in the soil for bulbs and grubs, leaving the ground pocked and heaved by morning. Nature's gardeners, their divots and tracks often fill up and hold water for days after it's rained, a small but significant hand in slowing the drying of the land.

A Question of Balance

Of course, left unchecked, the ungulates, the boar and the deer can become a costly menace to the farmer, particularly for those growing vines. Centuries of persecuting Italy's apex predator, the wolf, allowed ungulate numbers to swell far beyond what the land could comfortably bear. The Italian wolf (*Canis lupus italicus*) once ranged across the entire peninsula, feeding chiefly on deer and boar, its presence shaping both prey and forest. By the early 1970s, after decades of poisoning and persecution, barely a hundred wolves survived in the high forests of Abruzzo. With their disappearance, the balance collapsed: saplings were stripped, crops trampled, vines pilfered and hillsides eroded where herds of deer and wild boar grew too dense. The only control came from the hunter's rifle. In recent decades, however, the wolf has begun to reclaim its old dominion, moving silently down from the Apennines to reoccupy the woods and ravines from which it was driven. There are estimated to be about four thousand of them now, and the landscape of fear has returned, keeping the ungulates wary and on the move, never lingering too long in one place. Their howls mark the return of the original Holocene order, once again knitting itself back together.

As the great beasts faded from view, the smaller, quieter lives that had always been there grew newly visible. In the dark hollows of chestnut and oak, beetles endured where bison and bears had gone. The sky-blue longhorn *Rosalia alpina* still haunts the fallen trunks of Italy's ancient beeches, while in the forests of the Biellese Alps another survivor clings on – the iridescent *Carabus olympiae*, now restricted to a handful of valleys. There, among the beech and birch, an insect once nearly extinguished by deforestation continues to glitter in the shadows, a living emblem of resilience, and a reminder that even the smallest creatures can outlast empires.

For all the apparent abundance, not all forests are created equal. Much of Italy's woodland is still composed of *bosco ceduo*, coppiced stands cut and regrown on short rotations, once the backbone of the charcoal and firewood economy. These low woods, endlessly resprouting from the same ancient stools, supplied fuel for the furnaces of empire, glassworks and bread ovens for centuries. Today, they persist mainly through neglect: over-aged, shaded out and poor in structure, their tangled regrowth yielding little of real value beyond cheap firewood.

By contrast, true *fustaie* – high forests allowed to grow for generations – remain comparatively rare, yet they are the healthiest and most ecologically complete woodlands Italy possesses. Their tall, straight trunks provide the country's finest timber: spruce for violins and cellos, beech and maple for furniture, larch and fir for houses and bridges. Among the best examples is the Tarvisio Forest on the Slovenian border, one of Europe's oldest managed woodlands, where trees have been tended selectively for nearly a thousand years. Further west, in the Val di Fiemme and Paneveggio forests of Trentino, careful selection over centuries produced the famed 'resonance spruces' (almost invariably Norway spruce, *Picea abies*) prized by the luthiers of Cremona. Antonio Stradivari himself is said to have chosen his soundboards from these alpine valleys, where slow growth in the cold mountain air yields timber of exceptional density and tone. Together, these few *fustaie* show what a forest can be when patience and stewardship replace extraction – proof that the highest use of wood lies not in burning it, but in letting it grow.

Yet even the best-tended forests are proving vulnerable to the new realities of a warming, unsettled climate. In October 2018, the great windstorm Vaia tore through the Dolomites with hurricane force, flattening entire mountainsides of spruce and larch in a single night. More than 8 million cubic metres of timber were brought down – enough wood to build a city, wiping out centuries of careful silviculture. The Tarvisio Forest itself was struck, though spared the worst, while in the Fiemme Valley, cradle of the resonance spruce used by Stradivari for his violins, the slopes lay like a battlefield of splintered trunks. Across northern Italy, droughts now parch beech and fir on the southern slopes; fires creep higher into the hills each summer, and waves of alien pests turn whole stands to ash.

So apart from sanitation, medicine, education, wine, public order, irrigation, roads, freshwater systems and long stretches of peace, what did the Romans ever do for us? More than anything else, they left warnings about our interaction with the natural world: the costs of over-exploitation, the fragility of abundance and the ease with which prosperity erodes its own foundations. They may not have altered the global climate as we are doing today, but they almost certainly disrupted their own regional weather cycles. With forests and wetlands stripped away, the Mediterranean's capacity to hold moisture and summon rain was diminished, leaving an already arid land increasingly parched. Today, as climate change accelerates and fresh water dwindles, this broken water cycle is set to become one of the defining stresses for Italy and its neighbours.

CHAPTER SEVEN

Refuge

Nel mezzo del cammin di nostra vita
mi ritrovai per una selva oscura,
ché la diritta via era smarrita.
In the middle of the journey of our life,
I found myself in a dark wood,
for the straight way had been lost.

Dante Alighieri,
Divine Comedy: Inferno

Once a year, usually in May, a group of us meet in a wild, unfamiliar part of Europe, to hike, explore, reminisce and spend time together. The line-up is always slightly different, depending on availability. We are all well into middle age, married, some remarried, fathers, employed to differing degrees, greying around the edges, and, nowadays, mostly well-behaved. These annual excursions are a kind of outdoor men's therapy – an excuse to explore new landscapes and to safely share gossip or unspoken issues that might otherwise stay buried. Most would agree it's a healthy thing, especially our wives.

This year we are a group of four: myself and three of my oldest childhood friends. We have dispersed lives now, one in Portugal, one in Devon, another in London and me in Italy. We don't see each other nearly as much as we'd like, the complex logistics of our lives rarely allow for it, but they're the sort of friendships that don't ever seem to fade. One of them I haven't seen for more than ten years. After a warm embrace outside Fiumicino airport, the weight of absence quickly melted away and we were soon on our journey, grinding our way around the GRA, the orbital ring road of Rome.

I had been deliberately vague about the itinerary – a two-day hike on a portion of what's known as the Spirit Trail in the Maiella national park of Abruzzo, to visit some remote monasteries and hermitages in Italy's wild interior. I'd mentioned the possibility of seeing Italian grey wolves, Marsican brown bears and the flowering of Abruzzo's famed meadows. The Maiella, in particular, holds Italy's highest concentration of plant diversity – over a quarter of the national total – with around 140 endemic to the Apennines, several of which are found nowhere else. The forecast promised days of rain, but mountains make their own weather. We decided to trust that and go anyway.

What I hadn't told them was that, for me, the walk was also a kind of fieldwork, a way of tracing how Italy has always been a refuge, for animals, plants and people alike. Long before we came looking for solitude, these mountains had already offered it: first to bears, wolves and beech forests pushed south by the Ice Age; later to hermits, monks and early thinkers who, in their isolation, preserved fragments of the ancient world. I wanted to see how, in the centuries after empire, the wilderness of the Maiella had sheltered not just life, but ideas – the seeds of a new understanding of nature itself. In any case, as the host-organiser with not much prior knowledge, it's always best not to oversell the adventure or raise expectations too high, just in case it ends up a flop.

The first part of our journey took us around the southern outskirts of Rome. Off to the right, I pointed out the dormant Ice Age volcano of the Alban Hills, its crown veiled in cloud, once the quarry that built Rome and one day, perhaps, the force that might bury it again. After inching through the morning traffic, we finally broke free of the scrum and headed east toward Tivoli, where travertine, another of Rome's volcanic building blocks, is still quarried today.

We were now following a route that held quiet significance for what lay ahead. It was roughly the same path once taken by a young man from Umbria, about fifteen hundred years ago, a journey that would become one of the most consequential acts of spiritual retreat in Western history. Knowing my passengers were trapped for the next ninety minutes with no way of escape, I began my narration. My friends, to their credit, humoured me.

The Hermit's Tale

The young Benedict of Norcia arrived in Rome from Umbria around 500 CE to finish his studies. A serious young man, more devout than most, he came from the rural aristocracy, his father a senator and landowner of standing. Yet Benedict had been born into a world on the edge of collapse. The Rome he entered was no longer the city of marble his parents had imagined. Civil institutions persisted in form but not in life. The Senate still met but wielded no power. The Forum, once the beating heart of civic life, was choked with weeds and grazed by goats. The imperial baths stood dry; the great libraries, once the pride of Roman learning, lay neglected, their scrolls mouldering in silence. Corrupt officials presided over a city whose population had dwindled from a million to fewer than a hundred thousand.

Wandering through this faded grandeur, Benedict witnessed the slow unravelling of the Roman world. He soon realised that the education he had come to complete was no longer possible, and so, quietly, he turned his back on it all. With only his servant for company, he walked east into the hills, following the Via Tiburtina past Tivoli and tracing the Aniene River, the very artery that once fed Rome's aqueducts, into the rising Apennines. As our own car began to climb toward the mountains, I pointed out the turning for Subiaco, Benedict's eventual destination, and pressed on with the story.

For a time, Benedict stayed in the modest Roman town of Enfide (modern Affile), where he found fellowship among a small community of holy men. During this early sojourn, he became a figure of mild local wonder for repairing a broken *capisterium*, a clay grain-sorting vessel that everyone else had written off as useless. With miracles apparently in short supply, this bit of practical handiwork was hailed as his first sign of divine favour. Before long, people pressed him for more wonders and to fix more of their things. But this sort of provincial DIY miracle-working was not why Benedict had left Rome. So he departed again, plunging deeper into the wilds of the Simbruini Mountains in search of greater solitude and meaning.

In the Aniene valley, Benedict of Norcia was guided by a monk named Romanus who recognised in him a restless spiritual purpose. Romanus led him to a cave high in the limestone cliffs. Once again in Italy's history, a rough-hewn cave became a refuge in uncertain times. For the next three years, Benedict lived here as a hermit, sustained by Romanus' quiet care – food, water and the rare scrap of news from the world below – while he shaped the principles that would guide his life in service of God. Today, the cave, the Sacro Speco, is enveloped by a magnificent monastery complex, its walls covered in frescoes that trace the birth of Western monasticism from this improbable mountain cell.[1]

When Benedict finally stepped back into the outside world, the experiment was brief and almost disastrous. A community of monks, whose abbot had recently died, implored him to take charge. But when he imposed a strict regimen of rules and discipline, their enthusiasm quickly curdled. In a decidedly unholy act, they tried to poison his wine. Upon blessing his cup, it shattered in Benedict's hands, saving his life and revealing, with divine clarity, that his popularity had run its course. He fled from the mutinous brothers and returned once more to the silence of his cave.

Ora et Labora

The poisoned cup seems to have sharpened Benedict's mind. In a world that had clearly lost its moral bearings, how could men of faith like himself hope to live, work and pray together without descending into chaos? What was needed was structure, discipline and a daily rhythm, otherwise discontent would always fester, and poisonings – or worse – would follow.

When he emerged from the cave again, his purpose had hardened. He founded not one but twelve monasteries around Subiaco and the Aniene valley, each with an abbot presiding over twelve monks, bound by a strict devotion to prayer and work: *ora et labora*. Through trial and error, Benedict had come to realise that for most people renewal came not through solitude or extreme austerity but through communal labour. The hermit's life was for the rare few who could endure it. For everyone else, the way to a fulfilled life in God's service lay in rhythm and order: building, gardening, cleaning, reading, writing, praying at regular hours, teaching, sleeping soundly and generally spreading the good word to those around them.

His reputation quickly spread, and people flocked to the man who seemed able to create order wherever he went. His twin sister, Scholastica, soon joined him, founding the first convent for nuns nearby in the same spirit of prayer and work. But fame brings envy. A rival priest named Florentius began a campaign to discredit him and his followers, resorting to increasingly sordid tactics, at one point even marching naked girls through the monastery courtyards to tempt their vows of chastity, followed, inevitably, by another poisoning attempt.

Benedict knew it was time to move on. He headed further south through Lazio into the rugged uplands of Campania, where the mountains begin to lean toward the sea. On a prominent hill overlooking the Liri Valley, he came upon the crumbling remains of a pagan temple to Apollo, a place he instinctively recognised as ripe for renewal. This would become Monte Cassino, the most influential monastery in the Western world, the one that would define his legacy and crystallise the ideal of Christian monasticism.

It was here that Benedict composed his Rule – a concise, practical guide to communal life that would shape Europe's religious culture for centuries to come. Deceptively modest, it demanded neither extreme asceticism nor total renunciation of property, but balanced prayer, labour and rest within a framework of discipline and humility. It offered a middle path between isolation and indulgence, between chaos and control. The Rule of St Benedict became the template for nearly every order that followed, its spirit echoed in monasteries and hermitages from the Alps to Sicily. After the Benedictines came a new wave of reforming and mendicant orders: the Cluniacs and Cistercians; the contemplative Camaldolese and Vallombrosans; the austere Carthusians. By the thirteenth century, the great mendicant movements, the Franciscans, Dominicans and Augustinians, had carried the spirit of monasticism into the towns and cities. Yet many, the Franciscans above all, remained rooted in the countryside, carving hermitages into remote valleys and mountainsides, where nature was no longer merely a backdrop but an active participant in devotion.

As the Justinian Plague reached Rome around 541 CE, the great urban centres, once engines of commerce and civic life, had become prime targets for marauding armies and contagion. With the offices and institutions of the Roman world crumbling into irrelevance, monastic communities offered something few others could: safety, shelter and continuity, places that felt spiritually closer to God: mountaintop perches near the heavens, far from the chaos below.

The upper, austere landscape of Abruzzo revealed itself from within the protection of the car: forested lower slopes climbing to the tree line, above which bare, snow-laced summits dissolved into mist and cloud. In the distance, walled villages clung to their hilltops, each one bearing its legacy of refuge and abandonment. These were once the lands of the Marrucini and Peligni, and above all the Samnites – the fiercest of Rome's Italic foes, and the last to be subdued. A million years ago, herds of straight-tusked elephants would have roamed these valleys; ten million years ago, it was all a tropical sea. Ahead stretched the wide plain of Fucino, once the third-largest lake in Italy, now transformed into an immense agricultural patchwork, another testament to Roman ambition and ingenuity. For a few minutes, I let the silence settle, allowing my friends to absorb the mountainous world we had entered: the heart of the Apennines, to my mind one of the most significant landscapes in all of human history.

For millennia, these mountains have been more than a backdrop to history – they have been its bridge. Running the length of the peninsula, the Apennines form one of Europe's great faunal corridors, a living artery through which life still flows. Wolves, bears, deer and eagles still move along its ridges as they have since the Ice Age, passing from one refuge to the next across a chain of parks and protected forests that together form Italy's true wilderness spine. But after the fall of Rome, this ancient passageway took on a new purpose. The corridor that once channelled animals across the peninsula began to channel ideas, as monasteries took root deep in its folds and along its river valleys. The same transhumant routes – pathways first traced by Neolithic shepherds more than six thousand years ago and used ever since to move livestock between summer and winter pastures – that guided herds and flocks now carried hermits, pilgrims and scholars, linking distant valleys into a single breathing landscape and carrying with them the seeds of knowledge that would one day help reawaken the world.

The Knowledge

The monks carried with them their most vital possession: knowledge. Collections of scrolls containing the works of Cicero, Virgil, Ovid, Livy, Pliny the Elder, Seneca and Boethius were salvaged from the collapsing Roman world and secured within remote monastic libraries, where they were painstakingly copied and preserved in the silence of the scriptoria, the dedicated writing rooms of the monastery.*

Among the Christian theological texts being copied in the scriptoria were early works of natural science, medicine, art and cosmology: observations on plants, animals, minerals, pigments, the stars and the human body. Many of these ancient treatises later came to be viewed with suspicion, their pagan origins or speculative tone edging perilously close to heresy. Some were banned outright; others survived only by chance, reduced to a single manuscript left to moulder in a damp monastic library high in the mountains. Their survival was as fragile as that of the cave salamander or the owl moth: ancient lineages confined to hidden refuges, persisting in silence until the world was ready to rediscover them.

One text in particular, *De Rerum Natura* by Lucretius, was miraculously rediscovered in 1417 by the humanist Poggio Bracciolini in a Benedictine monastery at Fulda, in present-day Germany. Its heretical Epicurean vision of a universe governed not by gods but by atoms had been condemned by the Church and lost for centuries. Quietly, almost subversively, it re-entered circulation, its verses helping to ignite the first flames of humanism at the dawn of the Renaissance.

---

* Many of the great Greek thinkers – Plato, Aristotle, Homer, Galen and Theophrastus among them – were largely lost to the Latin West. Their works endured only through the scholarship of the Islamic world, where they were translated, commented upon and preserved. Centuries later, through the translation centres of Spain, Sicily and southern Italy, these texts returned to Europe, carrying with them layers of Arabic insight that helped ignite the intellectual renewal of the High Middle Ages and Renaissance.

Undoubtedly the most relevant, and the least imperilled, of all of the works copied in the scriptoria was Pliny the Elder's *Naturalis Historia*. Written in the first century CE, and the most complete of all surviving classical texts, it became one of the most frequently copied and widely circulated works of antiquity. The oldest surviving fragments of the *Natural History*, copied in Italy scarcely five centuries after his death, now rest in the libraries of Bamberg and Leiden, the frail beginnings of a manuscript tradition that would carry his vision of nature through the Dark Ages. By the seventh and eighth centuries, it was being copied regularly in monasteries across Italy, and by the ninth, from Monte Cassino to Corbie, monks were reproducing the text anew, securing its passage into the Renaissance.

Unlike many who came before him, Pliny did not seek the hidden causes or divine meanings behind nature; instead, he simply presented it as it was: a vast and wondrous mosaic of the living world. In its pages, elephants and eagles share space with herbs, minerals and clouds, each treated with the same curiosity and respect, as if all creation were of equal marvel.

Pliny was said to have moved through life in a state of near-permanent astonishment, and he endeavoured to convey this awe and reverence for the natural world in his writing, especially in what he observed most closely around him in Italy. He marvelled at its vines and springs, its olives and pines, its volcanic soils and temperate skies, declaring that, 'In the whole world, wherever the vault of heaven turns, there is no land so well adorned with all that wins Nature's crown as Italy.' He was the first great champion of the Italian landscape, praising it not through myth or divinity, but through observation and wonder. This state of astonishment was, fittingly, the same condition that led to his death: in 79 CE, unable to resist the pull of nature at its most terrible, he sailed across the Bay of Naples to observe the eruption of Vesuvius while it buried Pompeii, and never returned.

The *Natural History*'s encyclopaedic sweep, embracing cosmology, animals, plants, stones, medicines, geography, art and human invention, made it a foundational text of medieval learning. To monastic scribes, it was invaluable: a vast compendium of ancient knowledge that could be easily accommodated within the margins of faith. Pliny was an indefatigable collector and compiler claiming, with characteristic modesty, to have collected twenty thousand facts; modern counts suggest the number is closer to thirty-seven thousand. He was also remarkably diligent and transparent about his sources. Thanks to his habit of generous attribution, we know the names of many classical authors whose works would otherwise have vanished into obscurity. From the moment it was written, the *Natural History* stood as the most audacious attempt to catalogue the natural world in its bewildering entirety, and it remained Europe's primary reference on nature for more than fifteen centuries.

It is amazing to think that Pliny's observations remained authoritative for so long. Thanks to him, people believed for centuries that geese could hatch from driftwood and seashells, that bees might spontaneously arise from the carcass of a sacrificed ox, that ostriches could digest iron and that faraway tribes of humans possessed dog-like heads and barked instead of speaking. Fans of Harry Potter will be pleased to learn that it was through Pliny that the legend of the basilisk, the serpent so deadly it could kill with a single glance, first entered Western folklore.[2]

Despite its frequent absurdities, Pliny's *Natural History* also contains passages of genuine insight. His notes on soil erosion, medicinal plants, fossil seashells found far inland and the water cycle reveal a mind grappling earnestly with the forces of nature. His chapters on painting and pigments remain among our most important sources on classical art materials and techniques. Yet the *Natural History*'s true legacy lies not in its encyclopedic ambition, but in its spirit. Pliny is now often regarded as one of the world's first environmental thinkers, a voice whose warnings against deforestation, soil exhaustion and the reckless extraction of minerals and metals feel strikingly modern. His abiding sense of wonder and fascination with the living world reads today as an early expression of what Edward O. Wilson would later call biophilia: the innate human affinity for life and its systems. At a moment when we urgently need to recover a sense of reverence for the planet, there are far worse guides than Pliny the Elder, whose curiosity, humility and moral concern for Earth remain as inspiring as ever.

Survivalist Hermits

Unsure whether I still had my friends' attention after my Plinian ramblings, I announced that we were approaching Sulmona, the birthplace of Ovid, with the Morrone mountains rising starkly behind it – a key spiritual landmark for our journey. At their foot stands the Abbey of the Holy Spirit at Morrone, founded in the late thirteenth century as the headquarters of the Celestines: a reform branch of the Benedictines who embraced a more ascetic, vegetarian and often hermitic life. Their rule was laid down by the order's founder, Pietro da Morrone, later Pope Celestine V, remembered by history as 'the pope who quit'.[3]

Pietro da Morrone, who fasted every day of the week except Sunday, was a man who preferred the quiet life above all else. Drawn to solitude from an early age, he withdrew around 1230 to a cave high on Mount Morrone above Sulmona, where he lived as a hermit for many years. Like Benedict seven centuries before him, he soon attracted followers and a reputation for miracles. Yet where Benedict had recognised that most people needed the order and rhythm of communal life, Pietro sought the opposite – ever deeper solitude, stripping existence back to its barest essentials in pursuit of pure devotion to God.

As more followers gathered there, Mount Morrone gradually became too gentrified for Pietro's liking. After his ordination in Rome, he took a handful of disciples and vanished deeper into the mountains, into the wilds of the Maiella, a vast limestone massif and the uplifted bed of an ancient sea. It holds the Apennines' second-highest peak and remains the stronghold of wolves, Marsican brown bears, golden eagles, Apennine chamois and only the hardiest of shepherds. In its damp recesses lurk fire salamanders (*Salamandra salamandra*) and Apennine yellow-bellied toads (*Bombina pachypus*), while among the sunlit rocks lives the rare Orsini's viper (*Vipera ursinii*), a timid, grass-dwelling snake found only in a few scattered pockets of Europe. Vast beech forests mantle the lower slopes, rising steeply from the narrow, V-shaped valleys in a vivid, unbroken band of green. Where the trees abruptly end, the mountain reveals its bare skin, patchy rock and snow lying exposed to the cold, blustery winds.

This was our destination. As we rounded Popoli on the final stretch of the road, the great massif suddenly rose before us. I had brought my friends to see one of Europe's last true wildernesses. Nestled within its limestone folds, lay the place where Pietro found his ultimate spiritual retreat – the landscape in which he founded and restored the principal hermitages that became the heart of the Celestine order. Alongside Tibet and Cappadocia, the Maiella holds one of the highest concentrations of sacred sites in any mountain landscape. It was this long tradition of devotion, carved into caves and scattered across ridgelines, that led Petrarch to call it Domus Christi, the House of Christ.[4]

Res Herbaria

We awoke early the next morning and drove to Decontra, a small hamlet above Caramanico Terme, to begin our two-day trek. A local guide joined us, knowledgeable, patient and the perfect foil to my wandering commentary, occasionally expanding and grounding my ramblings with greater authority. For the moment, at least, the weather – and our luck – seemed to be holding.

We set out along one of the paths that meander across the open, treeless pastures. Climbing upward in the brilliance of the morning sun, we soon shed the outer layers that had seemed so necessary an hour earlier. With each step, the landscape gained in splendour, but our immediate attention was drawn downward, to the ground at our feet. Wildflowers were everywhere, bursting from the turf in dense, miniature constellations of colour.

Abruzzo is famed for its floral wealth and the Maiella stands at its heart. The park harbours more than 2,100 plant taxa, roughly a quarter of Italy's native flora, including between 140 and 200 species endemic to the central Apennines. Many are relics of the Ice Age that have survived here in isolation, evolving into unique mountain lineages. At least ten species, botanists believe, grow nowhere else on Earth. Among them are the Apennine edelweiss (*Leontopodium nivale* subsp. *alpinum*), a woolly, silver-white survivor of the glacial tundra that now blooms only on the park's highest ridges, and *Adonis distorta*, whose twisted pink petals open briefly between melting snowfields. Others include *Androsace mathildae*, endemic to the Abruzzo Apennines (known only from the Maiella and Gran Sasso massifs), and *Centaurea tenoreana*, a thistle confined to the limestone slopes of the Maiella itself. Each is a fragment of a lost world, remnants of the alpine flora that once covered much of southern Europe before retreating north with the ice.[5]

Feeling this might be the moment to slip back into my discourse, I pointed to a patch of yarrow beside the trail. I remarked how Roman soldiers had once packed its leaves into wounds to staunch bleeding, a practice later adopted and refined by medieval monks. Our guide nodded, adding the uses of another flower as an anti-inflammatory, and another to ease digestion. Monks, I explained, had long cultivated such plants in the shelter of cloistered gardens. Long before medicine became a science, the monastic apothecaries were among the first institutions to systematically observe and record the properties of plants, drawing on ancient authorities such as Theophrastus, Dioscorides, Galen and, of course, Pliny, and blending their teachings with local knowledge. From these gardens came the teas, salves and tinctures that formed the first foundations of pharmacology.

Medicinal gardens were once a defining feature of Italian monasteries. By the thirteenth century, the brothers of San Giovanni Evangelista in Parma were cultivating and processing healing herbs in a fully functioning garden and apothecary. At Monte Cassino, a modest *hortus medicus* supplied the infirmary with ingredients for salves and poultices, carefully noted alongside scripture in the scriptorium. In the forested quiet of Tuscany, the Camaldolese monks distilled their herbal remedies in a secluded pharmacy, a practice they maintain to this day.

Our guide breezily mentioned the Abbey of San Clemente a Casauria, here in Abruzzo, the birthplace of the region's formidable green liqueur, centerbe. Distilled from a hundred local plants, it could, under certain circumstances, be considered medicinal. In Florence, the Dominican friars of Santa Maria Novella founded their apothecary in 1221 to prepare balms for their infirmary. That institution, now the Officina Profumo-Farmaceutica di Santa Maria Novella, has grown into a global heritage brand with more than seventy-five boutiques worldwide. Its original shop in Florence remains the oldest continuously operating apothecary in existence, a living descendant of the medieval physic garden.

We paused beside a patch of purple elder-flowered orchids, their delicate bursts of colour flickering against the green. The Maiella harbours around eighty species and subspecies of wild orchid, more than half of those known in Italy, an astonishing concentration for such a rugged landscape.[6] Just beyond stood a little stone igloo, one of the dry-stone shepherds' shelters scattered across these slopes, tucked among the wiry shrubs of dwarf mountain pine (*Pinus mugo*). A fine white mist drifted over the hillside, veiling the view beyond, though for now the weather held.

Over time, the herbal wisdom of the monasteries began to spill out into the wider world. Rooted first in monastic and cathedral schools, the great intellectual revival of the eleventh and twelfth centuries saw learning migrate into the cities, where it reached a broader and more diverse body of scholars. The study of plants, once the province of cloistered herbalists, gradually became more systematic, laying the groundwork for modern science. The University of Bologna, founded in 1088 and often regarded as the world's oldest university, emerged as a key centre for transmitting and expanding medical and botanical knowledge.

Around the same time, the medical school of Salerno, widely regarded as Europe's first medical school, flourished as a cosmopolitan hub of medical instruction, blending monastic learning, classical texts and the new insights arriving from the Arab world, particularly through the writings of Avicenna (Ibn Sina), the Persian polymath whose medical encyclopaedia shaped European medicine for centuries. There, in the early fourteenth century, the physician Matteo Silvatico established Minerva's Garden, a teaching garden for medicinal plants that is often considered Europe's earliest botanical garden.

Two centuries later, the first formal university course on botany, *res herbaria*, the study of plants, was inaugurated at Padua in 1533. Soon after came the creation of the world's first university botanic gardens: in Pisa in 1544 and Padua the following year – a distinction still debated, though Padua's garden remains the oldest preserved in its original form. These were living laboratories where plants were cultivated, classified and exchanged, and in which the modern sciences of medicine, botany and pharmacology took root.

It was a former student of these monastic and medical traditions, Luca Ghini, who planted the first true seeds of botanical science. Trained as a physician and professor at the University of Bologna, he later moved to Pisa, where in 1544 he established a pioneering university garden that placed living plants at the centre of study and teaching. Equally revolutionary was his invention of the *hortus siccus*, the herbarium, a method of pressing and preserving plants so they could be studied, compared and taught throughout the year. This seemingly simple innovation transformed botany from a field practice into a systematic, scholarly science.[7]

Among Ghini's pupils was the brilliant Bolognese polymath Ulisse Aldrovandi, who carried his teacher's ideas further than anyone before him. Besides founding Bologna's own botanic garden, Aldrovandi assembled one of Europe's first great natural history museums – a vast cabinet of plants, animals, minerals, fossils (including fish fossils from Monte Bolca) and other marvels of nature. His collections were part theatre, part laboratory and part temple to curiosity. Pliny the Elder would have loved it. Long before Carl Linnaeus, Georges-Louis Leclerc, Comte de Buffon or Charles Darwin, Aldrovandi sought to gather the entire natural world beneath a single roof, establishing natural history as a discipline in its own right. He is also credited with being the first to use the term *geologia* – geology – in connection with the study of Earth.[8]

In the same century, a Sienese physician, Pietro Andrea Mattioli, rose to fame through his monumental *Discorsi* on *De Materia Medica*, a manual of medicinal plants originally compiled in the first century CE by Pedanius Dioscorides, a Greek physician serving in the Roman army.[9] Translating the text into Italian for the first time, Mattioli expanded it with detailed notes and descriptions of more than a hundred additional plants, many native to the Alps and Apennines. His editions were printed and reprinted across Europe, becoming the most influential medical-botanical text of the Renaissance. His Venetian publisher, Valgrisi, claimed to have sold 32,000 copies of the first edition.[10]

With each new printing, Mattioli's commentary grew more expansive, enriched by dried specimens and field notes sent to him by Ghini and Aldrovandi. When, in 1568, the work was finally adorned with the exquisite illustrations of Gherardo Cibo, himself a pupil of Ghini, the book achieved a kind of immortality: a perfect marriage of ancient wisdom and Renaissance artistry, uniting the careful eye of the botanist with the brush of the painter.[11]

I recounted to my friends how, not long ago, the librarian at the Biblioteca Alessandrina at La Sapienza University had kindly let me leaf through the world's only surviving edition of Mattioli's *Discorsi* hand-coloured by Gherardo Cibo himself. It is an extraordinarily beautiful object, one of the most exquisite fusions of art and science ever produced. Less convincing were some of the remedies within. Mattioli's proposed cure for drunkenness was to apply vinegar to the testicles, a procedure that, while probably hilarious when inebriated, is likely to be ineffective, if not mildly stimulating. I also strongly suspect that applying crushed raw crabs to your boobs is not the most effective treatment for post-partum breast inflammation. But I suppose you never know until you try.

With the mention of testicles and breasts, I managed to win back my friends' attention. Our upward climb paused beside a stone drinking trough fed by a nearby spring. In the rippling water, we spotted an Apennine yellow-bellied toad, one of Europe's most ancient faunal relics. The open pastures and flower meadows we had been enjoying were now giving way to a fringe of beech trees and our field of vision was about to change dramatically. Crossing the threshold of the forest, we entered a more perpendicular realm: a silvery-grey array of trunks, bright green new growth and an understory mottled with ochres and shadow. It was an abrupt sensorial shift, a clear delineation of a new phase in the journey. We plunged into the woods, where the path no longer ran straight.

Every so often, the beech forest broke open, the canopy parting to reveal the sky. Muddy white patches lingered across our path and we crunched through the last traces of sugary spring snow. Then, for the first time all day, our guide turned us downhill. The way grew more demanding: rocks and roots sprawled across the trail, forcing careful footwork and close attention. As we zigzagged lower, a feeling of arrival began to build. Rough stone steps appeared, deliberately but irregularly laid into the mountain's flank. Descending deeper still, we found handholds carved into the rock on our left; to the right, the trees thinned and the green slopes of the Maiella emerged, released at last from their shroud of cloud.

The final steps brought us to a sloped clearing pressed against a sheer wall of stone. A narrow staircase, carved directly into the cliff, climbed steeply to our left. At the top, an unguarded ledge hugged the rockface. We edged forward carefully, shoulders against the wall, the last stretch passable only on hands and knees. Two of our group, gripped by vertigo, turned back.

This was the entrance to San Giovanni all'Orfento, Pietro da Morrone's most cherished retreat, the least accessible and most remote of all his hermitages, cut directly into the limestone rock at 1,200 metres. He arrived here at around seventy years of age, accompanied only by a handful of devoted followers, and spent nearly a decade in disciplined solitude, suspended between wilderness and the divine. Below the hermitic caves that overhang the cliff, there had once been a gallery of cells and another chapel, now reduced to rubble and leaves. Excavations in 1995 uncovered Bronze Age remains, but scholars believe the caves may shelter far older histories still. Channels carved into the walls once gathered rainwater, feeding basins and a cistern – a feat of survival engineering that sustained life in this unforgiving place.

To imagine a man in his seventies surviving winters in this place, enduring snow, wind and hunger in little more than cassock and sandals, was staggering. Yet it was in extremity that Pietro found ecstasy. In early May the place was serene, warmed by spring sun and the fragrance of alpine herbs. We sat together on a great stone boulder before the complex, gazing out across the wooded expanse of the Maiella massif, one of the most entrancing views I have ever seen in Europe. According to legend, in his final winter here Pietro, frostbitten and near death, was carried down to Sulmona by his disciple Roberto da Salle. Within a year he was, much against his wishes, elected pope – only to resign five months later, in what became the most famous voluntary abdication in papal history, a precedent not repeated for personal reasons until Benedict XVI in 2013.

Winding Down

Governed by time and the feeling that our luck with the weather might be running out, we climbed our way out of the remarkable rock den and back into the beech forest. We retraced our original path and then, thankfully, began the long downhill stretch that would carry us through the afternoon.

The sound of water drew us on and the beech woods finally opened at the valley floor, where a clear, live, foaming stream wound its way suddenly beside us. We followed its course, flowing from bank to bank in the wide break of the beech canopy, knowing our destination was close. Thunder rolled along the surrounding peaks, announcing the storm's approach.

Emerging onto a narrow road, we climbed again just as the first drops of rain fell. Puffing hard as we rounded the final bend, we were drawn magnetically into Santo Spirito – a cream-coloured monastery complex that seems to grow organically from the wall of ancient, banded limestone that embraces it. Rock and masonry merge into one; trees, shrubs, flowers and lichens spring from every blotched crease and ledge. This was Pietro da Morrone's principal refuge – a place he first rebuilt and expanded in the 1240s, and which has thankfully been restored and renewed many times since. Jogging the last 50 metres, we reached its sanctuary just as the sky gave way. The rain, which had politely held off all day, now broke with its full and sudden weight, nature's holy spirit washing us into the church of the same name.

We spent the night here, entranced, well fed and reflecting deeply on all we had witnessed. Where we were – a fantasy carved by faith into mountains enveloped by forest – one feels the deep entanglement of humanity with the landscape. It is the perfect place to consider ourselves as humans driven by both belief and subservience to the rules of nature.

If they continue to be renewed, restored and protected, the great monasteries, abbeys, hermitages, sanctuaries and remains of wild rural Italy will continue to shape and preserve our human imagination as they have done for thousands of years. It is to them, and to the folds of the mountains that hold them, that we possibly owe our greatest tribute. In particular, it is in the Apennines, where we found ourselves, that Italy's role as the great refuge has endured: a corridor of life for millions of years, a stronghold for early hominids, shepherds and hermits, survivors of the classical world whose knowledge they preserved when the peninsula's first cities fell and the plains burned. It is these mountain sanctuaries that ensured not just survival but revival. Knowledge was gathered, ordered and rediscovered, and people learned again how to live within the limits and rhythms of the land. It is here that we learned how to begin again: in these high refuges, Italy's old world survived long enough for a new one to be born.

CHAPTER EIGHT

Revival

Divina natura dedit agros,
ars humana aedificavit urbes.
Divine nature gave us fields,
human skill built our cities.

Varro, *On Agriculture*

After what was probably one of the most peaceful nights of our lives, the next morning we set out on the final leg of our journey, toward Pietro da Morrone's other great hermitage: San Bartolomeo in Legio. The trek had taken on the feeling of a true pilgrimage, and for the grand finale, even the sun made an appearance.

As we walked, my thoughts shifted from the medieval world we had been tracing to a period far more familiar to me. For centuries, the monasteries and hermitages of Italy had served as refuges, safeguarding life, memory and fragments of learning while the lowlands endured upheaval. But by the time Pietro da Morrone became Pope Celestine V in 1294, a remarkable transformation was already underway. Italy, still politically fractured into republics, kingdoms and papal domains, had become the richest and most economically dynamic region in Europe. Its population, reduced to perhaps three or four million after Rome's fall, had rebounded to between ten and twelve million, making the peninsula once again one of the most densely populated places on the continent.

This resurgence was not a miracle of politics so much as one of geography and nature. Fertile plains, navigable rivers, defensible coasts, Alpine passes and a necklace of industrious cities allowed Italy to knit itself back into the wider world. Its mercantile republics – Venice, Genoa, Pisa and Florence – controlled Mediterranean trade; its bankers pioneered the instruments of international finance; its artisans transformed wool, silk and pigments into the luxury goods that underwrote Italian power. From the thirteenth century onward, towns that had once seemed vulnerable outposts became cultural powerhouses, capable of remaking Europe's imagination and helping to precipitate the Renaissance.

Walking on Water

Venice in particular, that most extraordinary of refuges, had been built upon timber piles in the saltwater marshes of the lagoon – a city literally wrested from the water, a defiant reinvention of nature in the face of the Lombard invasions of the sixth and seventh centuries. Geologically, Venice is a young creation, an afterthought of the Ice Age, built upon layers of silt that once lay far inland. Its setting is also a masterpiece of adaptation: a metropolis that rose not despite the water, but because of it. The Venetian Lagoon, a shallow, shifting mosaic of mudflats, saltmarsh and tidal channels, became both shield and sustainer, filtering waste, softening floods and providing fish, fowl and reeds for fuel and building. Twice a day, the tides redraw its boundaries, a landscape in perpetual negotiation with the sea.

When the German polymath Johann Wolfgang von Goethe first arrived in Venice in 1786, he called it a *Biberstadt*, a 'beaver city', a place where humans, like their animal counterparts, had learned to build with the water rather than against it. The foundations of the great city were laid in wood: millions of piles, mostly larch, alder and oak, driven deep into the lagoon mud, felled from forests across the Veneto, Friuli and Istria, in what is now Croatia. Among the most heavily exploited were the upland woods of Tarvisio, Carnia and Cadore, whose trees were cut and sent downstream to keep the city afloat. The stones came from Istrian quarries resistant to salt as well as the volcanic tuffs and trachytes of the Colli Euganei – dense, hard lava stones formed by ancient eruptions close to Padua and prized for their durability in a damp, corrosive environment. The mosaics glinted with minerals carried from distant lands; the glass too, born of sand, soda and fire, turned the chemistry of the planet itself into art.

Yet the lagoon remains a world apart, a brackish borderland where salt and fresh water meet, creating one of Italy's richest and most fragile ecosystems. More than a hundred species of fish and over two hundred species of birds inhabit its shifting channels, sandbars and reedbeds, from marsh harriers and shelducks to the elegant redshanks that stitch their way across the mudflats. For migratory birds, the lagoon is a sanctuary, a watery crossroads where Africa meets Europe. Every spring, flocks of flamingos, stilts and avocets return to these shallows, tracing routes older than the city itself.

Its most famous resident is the *moeche*, the soft-shelled lagoon crab, harvested in spring and autumn at the brief moment between moults – a Venetian delicacy whose survival still depends on the lagoon's delicate timing of tides and temperature. But new arrivals have changed the balance. The Atlantic blue crab (*Callinectes sapidus*), introduced accidentally through ships' ballast waters, now prowls the shallows in growing numbers, devouring clams, fish and even the native *moeche*. Like so many of Italy's modern invaders, it is both symptom and symbol – a creature thriving on the global interconnectedness we have created.

The lagoon, once rich in its own peculiar life, now mirrors the fate of Italy's coasts: blue crabs from America, Asian algae drifting up the Adriatic, and the same gulls and pigeons that haunt every harbour from Venice to Palermo. Recent studies identify the lagoon as Italy's foremost hotspot for alien macroalgae, home to more than thirty introduced species, with *Gracilaria* alone accounting for nearly half the total invasive biomass. The newest arrival, the red seaweed *Gracilaria vermiculophylla*, a prolific agar-producing alga, spreads in dense, suffocating mats that smother native seagrass and choke the shallows, transforming the lagoon's colours and textures as surely as tourism has changed its streets. Yet marine biologists studying the lagoon have noted that *Gracilaria*, though invasive, has paradoxically helped stabilise parts of this fragile ecosystem, outcompeting the green algae whose sudden die-offs once triggered anoxic crises. Even here, invasion can sometimes blur the line between loss and adaptation.[1]

Ars Naturae

By the fourteenth century, Venice straddled East and West, growing rich on its maritime empire and on its commerce with Byzantium and the Islamic world. It became Europe's ultimate entrepôt, channeling the wonders of the wider world – textiles, spices, plants and minerals – into the heart of the continent.

From this intermingling of commerce and ecology, of natural resources and human ingenuity, something remarkable emerged. The wealth that flowed through Venice and its rival republics seeded a new class of patrons and thinkers, igniting what was perhaps the greatest cultural reawakening the world has ever seen. The convergence of economic power, international connection and a hunger for secular knowledge became the soil in which the Renaissance took root. Monks had safeguarded memory; now merchants would fund its reawakening.

Painting, sculpture and architecture, once the anonymous work of craftsmen, monks and stonemasons, now increasingly came from the hands of celebrated individuals, eagerly sought by powerful patrons. The materials of Italy's landscapes – its marble, minerals and light – became the raw matter of a new visual language. In painting, the landscape itself began to emerge as a subject in its own right. Its mountains, valleys, rivers and forests provided the settings for visions of paradise, biblical dramas, portraits of princes and donors, and mythological epics. From Giotto onward, artists strove for startling fidelity, rendering familiar places, animals and people with a new naturalism. Combined with sacred narratives and classical mythology, whether in monumental frescoes or intimate devotional panels, this realism created an art that reshaped Christianity and antiquity alike. The interplay between Italy's physical terrain and the imagination of its artists would go on to define the Western conception of landscape for centuries to come.

On our walk down the Orfento valley, our guide pointed out caves that had once been mined for bitumen, long prized as both sealant and pigment binder. From the late medieval period onward, Italy's geology and mineral wealth once again became vital to artistic production, in a way not seen since Roman times. The rediscovery of nature was not only intellectual but material: artists, monks and alchemists learned to draw the very colours of their visions from the earth itself. Just as early scholars had distilled healing compounds from herbs, so too did they learn to derive colour from roots, minerals, insects and even moulds – substances that might become remedy or pigment depending only on how they were prepared.

As for so many other subjects, Pliny the Elder's *Naturalis Historia* remained the essential reference. His detailed accounts of minerals, ores and coloured earths, complete with instructions on how to extract and refine them, were studied and recopied, forming the backbone of artistic and alchemical practice throughout the Renaissance. Mining operations fanned out across the peninsula not only for ores and precious metals but for colour itself. *Terre verte* was quarried from Mount Baldo in the Veneto; red and yellow ochres from the iron-rich hills of Tuscany and the Veneto; malachite from Sardinia and Calabria; and cinnabar, the mercury ore used for the brilliant but poisonous vermilion, from the Tuscan magmatic belt. The sienna earths came from the clays around Siena; azurite from Tuscan copper deposits; and the deepest blacks from charred vines, bones, lamp soot and manganese-rich soils of central Italy. Lead white – *biacca* – produced by corroding strips of lead, gave the most durable white for tempera and oil, while lime white from Carrara and Verona limestone was favoured in fresco. Gold was panned from alpine streams near Monte Rosa and silver was dug from the open pits of Cugnano in the Maremma.[2]

Given its unrivalled access to global trade, Venice became home to the most dazzling pigments and the most celebrated colourists. The workshops of Bellini, followed by Giorgione, Titian, Veronese and Tintoretto, revelled in their richness. Specialist pigment shops, or *vendecolori*, supplied both artists and artisans with imported lapis lazuli, Mexican cochineal and exotic dyes. Florence, too, had its colour alchemists: the Dominican friars of the Gesuati at San Giusto alle Mura, famed for producing pigments for Ghirlandaio, Leonardo and Michelangelo. In May 1508, Michelangelo wrote from Rome to Fra Jacopo di Francesco of the Gesuati, requesting samples of blue pigment and a quantity of fine-quality azure: 'I have to have certain things painted here.' Certain things, in this case, being the ceiling of the Sistine Chapel.[3]

In sculpture and architecture, Italy drew once again upon the elemental wealth of its own mountains. Carrara marble, luminous and fine-grained, had been quarried in the Apuan Alps since Roman times and became the stone of choice for Renaissance masters. Michelangelo himself travelled to the quarries to select his blocks for the David and the Pietà; a century later, Gian Lorenzo Bernini would do the same for his *Apollo and Daphne* and *Ecstasy of St Teresa*.

Across Italy, cathedrals and civic buildings gleamed with the same geological brilliance: deep green serpentine from the Apennines near Prato, warm red marble from Verona and white Carrara from the north – Italy's later national colours written in stone. Together, they formed the polychrome facades and striped pavements of Florence's Duomo and Baptistery, Siena's cathedral and Orvieto's radiant frontispiece. The churches of the Renaissance thus embodied not only faith but geology: the colours of the earth reimagined as the hues of heaven.

Eyes of a Lynx

As we walked, I thought about our merry little band and decided to finish with a story of another group of four friends who, four centuries earlier, embarked on a very different kind of pilgrimage. Their shared adventure helped ignite one of the most important turning points in the history of science. It marked the culmination of Italy's surge in natural observation, the birth of modern botany and the flowering of artistic brilliance, all coming together in Rome at the dawn of the seventeenth century. For a brief but astonishing moment, religion, scientific curiosity and artistic imagination merged in a way rarely seen before or since.

In Rome in 1603, the eighteen-year-old nobleman Prince Federico Cesi, together with three like-minded friends, founded the Accademia dei Lincei, or the Academy of the Lynxes, named for the animal's fabled sharp sight, which they aspired to emulate in their study of the natural world. Often considered the world's first modern scientific academy, it was dedicated to direct observation, free from fear, repression or slavish adherence to Aristotelian authority, which had prescribed knowledge as fixed and emanating from authority and tradition, rather than arising from direct observation. Cesi drew around him polymaths, botanists, physicians, mathematicians and astronomers, united by a restless spirit of inquiry. Their statutes denounced marriage as a distraction from study and explicitly barred monks from membership, seeking to protect the academy's independence from both domestic obligation and ecclesiastical control.

When Galileo Galilei joined in 1611, he became its intellectual heart. Signing himself Galileo Galilei Linceo, he dedicated to them his work on sunspots, which offered fresh evidence for a dynamic, imperfect cosmos, contradicting the Church-sanctioned belief in an unchanging heaven. As Galileo's observations lent weight to Nicolaus Copernicus's heliocentric theory, the Lynxes became his most loyal defenders, even as his discoveries began to draw the wary attention of the Inquisition.

The Lynxes began to amass a remarkable collection of specimens, drawings and studies, attempting nothing less than to document and dissect the natural world in its entirety. Above all, they were pioneering field scientists. Federico Cesi's observations of fossilised wood around his family's estate at Acquasparta in Umbria count among the first detailed palaeontological studies ever undertaken, recognising in petrified trunks the vestiges of ancient forests. Their artists also produced what may be the earliest images ever drawn with the aid of a microscope, a term coined by their member Johannes Faber after examining Galileo's newly improved lenses. Using these instruments, the Lynxes recorded magnified views of bees' wings, plant fibres, insect eyes and mould spores with extraordinary accuracy. These drawings, collected in Cesi's *Tabulae Phytosophicae* (1628) and in Francesco Stelluti's illustrated edition of the *Apiarium* (1625), reveal an almost shocking level of detail: the first glimpses of the microcosm ever rendered by human hand.

Emerging from the final stretch of beech forest, we stepped into open meadows, once again blanketed with wildflowers, shepherds' stone igloos and scattered shrubs under a sky suddenly cleared to blue. Ahead, the valley broadened, its rocky ridgelines layering bands of green and grey.

'Spot the hermitage,' our guide challenged.

We scanned the slopes, seeing only rock, pasture and trees. Nothing revealed itself.

'Lower,' he said.

Only then, with the binoculars trained carefully, did the building resolve: a faint geometry of walls, a bell tower and a stair etched into the slope, all but swallowed by the rock around it. This was San Bartolomeo in Legio, Pietro da Morrone's most visually arresting refuge, a hermitage he rebuilt in 1250 that is so hidden it seems less constructed than conjured from the mountain itself. Nestled within the curve of a limestone ledge above the Orfento River and reached only by stairs cut directly into the stone, it appears and disappears with the changing light, as if the hillside were reluctant to give it up.

We looked on in admiration not only at the monks' perseverance in carving out such an impossible dwelling, but at their unwavering devotion to a life lived in dialogue with the landscape. In its own way, this remote sanctuary was as much a temple to observation as the Academy of the Lynxes, both born from the same Italian instinct to look closely at nature, to find meaning in its smallest details and grandeur in its vastness.

The Academy of the Lynxes did not long outlive Federico Cesi, collapsing soon after his premature death in 1630. Yet its spirit endured in one of its members, Cassiano dal Pozzo – Roman antiquarian, collector and fellow Lynx – who absorbed the academy's drawings and natural studies into his vast *Museo Cartaceo*, or *Paper Museum*. Over the following decades he commissioned thousands of watercolours, prints and drawings, creating one of the most extraordinary visual encyclopaedias of the seventeenth century: botany, zoology, fossils, architecture, textiles, antiquities.

Though formally only a cardinal's secretary, Cassiano was one of Rome's most influential patrons, supporting artists such as Bernini, Nicolas Poussin and Pietro da Cortona. His palazzo on Via dei Chiavari became a living Wunderkammer, filled with paintings, antiquities and exotic animals – a baroque echo of the Lynxian dream to observe and record the whole of nature. Then, like so many forests and manuscripts that had vanished in Italy's long history, the *Paper Museum* disappeared, until in 1986 the art historian David Freedberg opened a cupboard at Windsor Castle and found an enormous cache of drawings and watercolours: around 2,700 in total. It was the natural history portion of Cassiano's dispersed collection, acquired by George III in 1762 and forgotten for nearly two centuries. Further research traced other volumes to London, Paris and Montpellier, and in 2023 the final volume of the Cassiano dal Pozzo *Paper Museum Catalogue Raisonné* was published, completing one of the most remarkable scholarly restorations of modern times.

'Have you seen it?' asked one of my friends.

'Yes,' I said. 'The Accademia dei Lincei in Rome has a copy.'

'But didn't the Lynxes collapse after Cesi's death?'

'They did, but they were revived in the nineteenth century. And they're thriving today.'

Founded anew in 1874, the modern Accademia Nazionale dei Lincei is Italy's national academy of science and humanities, and one of the oldest scientific institutions in the world. Its members have included Galileo, Charles Darwin, Albert Einstein and Enrico Fermi. Housed in the sixteenth-century Palazzo Corsini in Trastevere, it continues to foster the study of both nature and culture, an unbroken thread from the first Lynxes' microscopes to the satellites and climate models of our own age. In a country where knowledge and beauty have always shared the same soil, the Lincei stand as the clearest proof that curiosity itself can be a form of faith.

The prevailing narrative often casts the Church as science's natural enemy – resisting Copernicus, silencing Galileo, recoiling at Darwin. It was certainly the version I absorbed as a student. Yet the truth is more intricate, and more Italian. The same institution that condemned Galileo also gave him his telescope; the same faith that feared the stars charted them on monastery roofs. Across the centuries, the Church did sometimes try to restrain inquiry, but it also sponsored and preserved it, safeguarding ancient texts, founding Europe's first universities and fostering religious orders that mapped the heavens, measured the earth and kept watch over its seasons. Many of Italy's greatest scientists – Galileo, Evangelista Torricelli, Athanasius Kircher, Angelo Secchi – worked not in defiance of faith, but within it.

The dialogue between faith and reason has never been a straight line, but a winding path, like those we had just walked, sometimes leading to revelation, sometimes to ruin and always back to the same enduring question: how to understand the world without diminishing its wonder.

Even the spirit of the great monasteries, today with sadly dwindling members, continues to exert a quiet but lasting influence. Italy's landscapes still offer refuge to those seeking a different rhythm of life. Across the country, dozens of eco-villages, communes and so-called intentional communities occupy the same hills, valleys and forests that once sheltered hermits and monks. Some are informal communes of squatters; others have a more spiritual or ecological focus, or are simply groups of like-minded friends. Yet all share a desire to live beyond the routines and pressures of modern society.

In Tuscany's Garfagnana, for example, the Valle degli Elfi has, since the 1980s, been home to families committed to self-sufficiency, communal living and a minimal ecological footprint. In Piedmont, Damanhur, founded in the 1970s in the foothills of the Alps, has grown into a federation of underground temples and communities devoted to art, spirituality and sustainable living. In Maremma, Nomadelfia continues the Catholic communitarian experiment begun by Don Zeno Saltini after the war, a living extension of Benedict's idea that faith and work are indivisible. Scattered through Abruzzo, Calabria and the Alps are similar projects: people who, like the monks before them, see in Italy's landscapes a chance to carve out a new way of being, less tethered to the economic mainstream and more attuned to nature's cycles.

It is a continuity that runs deep in the country's story: from the monks of the Apennines to today's *nuovi eremiti* – the new hermits who retreat into vans, cabins or off-grid communities; from Benedict's order to the eco-villages, rewilders and regenerative farmers of our own age. Each has sought, in their own way, to reconcile the material and the spiritual, to live within the limits of place rather than above them.

The Hiker's Tale

Having arrived at San Bartolomeo in Legio from the river below, we now took the stairs heading upward, along a path thick with bright red valerian and great flowering cistus shrubs. As the trail lifted us onto an open terrace of the mountain, the green plains of Abruzzo unfurled across the horizon. The distant tinkling of bells brought our eyes to a flock of sheep drifting slowly across the meadow, escorted by two white Maremma sheepdogs and the dark, upright figure of the shepherd with his long staff.

It was a scene as old as Italy itself. The slow, seasonal migrations of shepherds and their flocks along ancient mountain paths – the tradition known as *transumanza* – have shaped the rhythm of life in the Apennines for at least five millennia. Every spring and autumn, livestock still travel between the lowland pastures and the high summer meadows, a practice with roots deep in the Neolithic world. In an almost unbroken chain, generations of Italian shepherds have witnessed the transformation of the land around them: the retreat of forests, the spread of fields and vineyards, the rise of monasteries, villages, towns and, eventually, modern cities.

As the flock moved across the meadow, it struck me how fragile this ancient rhythm has become. The paths of *transumanza* still thread through the Apennines, but the world around them has thinned. Across Italy's interior, the pastoral landscapes that shaped its history are emptying at a pace unseen since the Middle Ages.

Today, abandonment has hollowed out much of the country: more than six thousand settlements stand empty, and tens of thousands of villages have seen their populations collapse, many losing over 90 per cent of their inhabitants – most of them scattered along this same mountain chain. The slow dissolution is visible everywhere: shuttered schools, collapsing roofs, terraces surrendering to scrub. A few headline-grabbing schemes selling houses for one euro may lure adventurous newcomers, but they are no substitute for a long-term strategy. Without a revitalised and diversified rural economy, the exodus will continue; villages will fall silent; and Italy's birth rate, already among the lowest in the world, will sink still further. None of this should surprise us: more and more people are crammed into small urban apartments, without the space, stability or confidence to start families.

Italy's population today hovers around fifty-nine million, an astonishing figure when placed against deep time: for roughly 97 per cent of the period Homo sapiens has lived in the peninsula, the population likely never exceeded a hundred thousand. For tens of millennia, survival was marginal, with humans clinging on through the severest phases of the Ice Age. Only in the past two centuries did numbers rise sharply, from seventeen million to nearly sixty million. Yet that brief surge has now reversed. Since peaking in 2014, Italy's population has begun to contract; without immigration, it could halve by century's end.

And yet movement, renewal and exchange have always been our species' saving graces. Italy – poised at the meeting point of Europe, Africa and Asia – has long been a crossroads of that human story. Its people today carry the highest genetic diversity in Europe, the imprint of tens of thousands of years of migrations and encounters. It would be a tragedy, not just for Italy but for humanity, if that inheritance were allowed to narrow and fade. Rewilding the land must go hand in hand with rewilding our presence within it: restoring not only habitats but the living communities that give them meaning.

Despite this, even as the villages empty and the old rhythms falter, a countermovement is quietly gathering force. Across Italy, a new generation is rediscovering the interior not as a relic of the past but as a frontier of possibility. If the shepherds once traced these ridgelines in search of pasture, today others follow them in search of meaning, connection and belonging – the beginnings, perhaps, of a rural renaissance.

Few have embodied this shift more vividly than the young collective Va' Sentiero. Between 2019 and 2021, they undertook a feat that borders on the mythic: walking and documenting the entirety of the Sentiero Italia, an 8,000-kilometre route that threads from the Slovenian border to Sardinia, one of the longest hiking trails on Earth. Starting in Trieste, they followed the Alpine arc from east to west, rounded the Maritime Alps and then traced the full spine of the Apennines from Liguria down to Calabria, before crossing to Sicily and ending in Sardinia's rugged uplands.

What began as a small expedition became a kind of national procession. Along the way they photographed, filmed, mapped and recorded the stories of the landscapes and communities they encountered, drawing attention to places that had slipped from Italy's imagination. More than three thousand people joined them for stretches of the journey – a scene at times reminiscent of *Forrest Gump*, a growing human thread weaving its way through abandoned hamlets, high pastures and forgotten valleys.

Their walk did more than revive an obscure mega-trail. It revealed, with extraordinary clarity, the scale, richness and untapped potential of Italy's wild interior: the cultural patrimony embedded in old stone terraces and shepherds' huts; the biodiversity sheltered in Alpine cirques, karst plateaus and Apennine forests; the deep continuity between landscape and identity that persists even where human settlement has retreated. Above all, it underscored the urgency of protecting these places – and the possibility of renewing them. Va' Sentiero showed that Italy's future need not be confined to its crowded coasts and cities. The mountains, long treated as margins, may once again become sites of imagination, enterprise and return.

However, just as new walkers are reclaiming the mountains, a different kind of return is unfolding within the streets of these abandoned towns. An innovative form of hospitality, born in Italy and now spreading internationally, is quietly stitching life back into places that once seemed destined to fade. The concept draws its inspiration from the reconstruction that followed the devastating Friuli earthquake of 1976, in Italy's northeast frontier, when planners began imagining ways to revive damaged villages without emptying them of their residents. Two decades later, this philosophy took formal shape as the *albergo diffuso* – the 'scattered hotel' – with the opening of the first official example in Bosa, Sardinia, in 1998. From there, the model spread across the peninsula.

In 1999, the medieval village of Santo Stefano di Sessanio in Abruzzo, near the crystalline waters of the Tirino River and across the valley from where we were standing, was almost entirely abandoned, its houses crumbling into ruin. Six years later, after a meticulous and sensitive restoration, the *albergo diffuso* Sextantio opened with guest rooms dispersed throughout the village in restored stone houses linked by cobbled lanes and archways. It is one of the most enchanting places I have ever stayed. Every detail, from limewashed walls to handwoven blankets, was guided by archival photographs and local memory.

The project revived not only architecture, but traditional crafts, foodways and livelihoods, breathing life into a place on the brink of disappearance. Though Sextantio is now a sought-after luxury destination, it catalysed new hotels, restaurants and small businesses, ensuring the village's renaissance continues. Today, there are around 150 scattered hotels across Italy, spread through medieval *borghi*, fortified hill towns and sprawling country farms.

Our own journey was now drawing to its close, and our final stop was, fittingly, a small medieval church dedicated to Thomas Becket, built only decades after his martyrdom, on the outskirts of Caramanico Terme. With that, the Chaucerian framing device of our hike felt complete. I hoped that the tale I had recounted had kept spirits up, as Chaucer intended his pilgrims' stories to do. Then we turned back toward Rome and, soon after, to our separate lives, each of us, I like to think, carrying a sharper sense of how landscapes shape us and how we, in turn, shape them.

CHAPTER NINE

Abundance

O fortunatos nimium, sua si bona norint, agricolas.
O greatly fortunate farmers, if only they knew their blessings.

Virgil, *Georgics*

Each spring, between February and April, the morning dog walk shifts from routine exercise into a slow, careful, almost meditative patrol. I start scanning the terrace edges for one thing above all: wild asparagus. It's the moment I look forward to most each year – those thin green and purple spears erupting from the soil are one of the great joys of the food calendar and the first true sign that winter is loosening its grip.

It grows everywhere around us: along the margins of fields, at the base of olive trees, in the hedgerows, and later in the season, deeper into the woods. It seems to define the boundary between cultivated and uncultivated land, a quiet marker of the wild fringe. Once your eye adjusts, hunting it becomes an obsession. Time ebbs away, and one often finds oneself deep in an unknown part of the woodland without realising how you got there.

For a couple of months, depending on rainfall, I'm foraging and eating wild asparagus almost continuously. The thin little spears belie their intense flavour, more aromatic, more earthy and pungent than the cultivated forms. I cook them in omelettes, risottos, pasta dishes and sometimes just roasted quickly with a bit of oil. They appear precisely at the moment when my winter vegetables are on their last legs and, more than anything else, they mark the start of the new vegetative year. Over time, I have learned as much when to leave them in the ground as when to harvest them – taking only what is strictly ready and necessary, and leaving the next generation for another year.

Native to the Mediterranean basin, wild asparagus has likely been gathered since Palaeolithic times. We know the Egyptians and Greeks ate it, but it was the Romans who truly revered it. They were ravenous consumers, believing it to be both a health-giving diuretic and an aphrodisiac. The Romans considered anything vaguely phallic to be an aphrodisiac. Cato the Elder includes advice on its cultivation in *De Agri Cultura* , and it is later mentioned by Varro, Columella, Pliny the Elder and even Apicius in his cookbook *De Re Coquinaria*. Such was the Roman appetite that they were the first to domesticate the plant, establishing large asparagus plantations, especially around Ravenna, and even constructing special ships to transport it swiftly to Rome.[1]

Besides asparagus, the Romans were enthusiastic domesticators and refiners of many edible plants. Over the centuries, they bred and developed varieties of celery, leek, fennel, garlic, cardoon and lettuce. Parsley (*Petroselinum crispum*) is considered native to Sardinia, from where it was first domesticated, quickly becoming the world's favourite garnish. One of their most important botanical legacies was the cultivation of wild cabbage (*Brassica oleracea*), a hardy, salt-tolerant plant that grew naturally along the marshy, coastal zones of the Mediterranean. They began selectively breeding it into more palatable and productive forms, giving rise to early versions of kale and eventually laying the groundwork for broccoli and cauliflower. Among its most vocal admirers was Pliny the Elder, who praised cabbage for its near-magical medicinal properties. He cited Cato the Elder's recommendation that the urine of a person fed exclusively on cabbage should be collected, warmed and applied to relieve ailments of the joints. 'Little boys,' he added, 'if you bathe them with such urine, never become weak.'[2] I suppose they can be thankful it wasn't from someone dieting on asparagus.

Green Gold

Over the years of getting to know my little farm in the Maremma and its surrounding margins, I've come to understand that the plants growing here tell a long and continuous story of food in Italy. Half of our land is classic Mediterranean scrub and forest – a patchwork of resilient, hard woods and aromatic species that have held their ground for thousands of years. I'm almost certain a Roman farm once stood where ours does today, and it's highly likely the Etruscans were earlier custodians of the land. In the steep woods that fall away from our plain, I've found remains of old stone terracing – traces of a time when cultivation here was far more extensive than the five hundred or so olive trees we now tend on the other half of the land.

Those olive trees are themselves the product of thousands of years of gradual domestication, a process that likely began in the Fertile Crescent some seven thousand years ago. All modern olives, whether grown in Tuscany, Puglia, Andalusia or Tunisia, descend from the wild olive, *Olea europaea* var. *sylvestris*. These hardy shrubs still grow spontaneously in the maquis around our farm, particularly along stony slopes and field margins. Their fruit is small and fiercely bitter, but dense with oil, a genetic reservoir that early farmers across the Mediterranean began to cultivate and refine more than six millennia ago.

A few kilometres from here stands the Olivo della Strega – the Witch's Olive – of Magliano in Toscana, believed to be more than three thousand years old. Its great hollow trunk, blackened by time, still sends up shoots that bear fruit each autumn. Further afield, in Sardinia, another patriarch endures: the 4,000-year-old S'Ozzastru, whose gnarled limbs still spread above the Gallura plain. Thought to be the oldest olive in Europe, it too bears fruit, a reminder that the story of domestication is not one of replacement but of continuity, where the ancient and the cultivated still share the same root.[3]

On our land, we grow three main cultivars: Frantoio, Leccino and Moraiolo. We harvest and blend them together usually in mid-October, producing a green, peppery oil packed with polyphenols, which, like any true olive oil grower, I will proudly claim is the best there is. These varieties, staples of central Italy, reflect centuries of selection and adaptation to Tuscany's rugged soils and changing climate.

Alongside them are more than five hundred known cultivars of olive grown across Italy, by far the greatest diversity of any country on Earth. Spain, the world's largest producer, has barely half that number. This astonishing variety tells a deeper story. Italy sits at the very heart of the olive's evolutionary journey: one of the Mediterranean's great secondary domestication zones, where early farmers and traders – Phoenicians, Greeks and Etruscans – brought cultivated strains that mingled and hybridised with the hardy wild olives already rooted in Italian soil. Over thousands of years, the peninsula's fractured geology, shifting microclimates and isolated valleys did the rest, shaping distinct regional lineages as surely as they shaped its dialects and peoples. Each valley, hill and coastal terrace produced its own answer to the same ancient tree. To grasp this is to glimpse the secret of Italy's abundance. Its unrivalled agricultural diversity, like its cultural one, was born from restless geography and from a long, intimate conversation between cultivation, wildness and resilience.

Yet even this ancient resilience is now under siege. A single bacterium, *Xylella fastidiosa*, has laid waste to vast tracts of olive groves in Puglia, blocking the vessels that carry water and nutrients and leaving once-living trees grey and skeletal. Despite heroic and often traumatic efforts at containment, there is still no cure. Since 2013, when it was first identified in the southernmost province of Puglia, the pathogen has killed more than twenty-one million olive trees, about a third of the sixty million that are cultivated in the region. Native to the Americas, *Xylella* is spread from plant to plant by the meadow spittlebug (*Philaenus spumarius*). And olives are not its only victims: almonds, stone fruits, grapevines, citrus, lavender and even native oaks and elms can be infected. The evidence suggests it is slowly, stubbornly spreading north.[4]

Wild

The other half of the farm tells the wild, foraged story of Italy's native food as well as traces of its forgotten fruits. Not that the two areas are entirely distinct. Unlike the more modern, intensive plantations trained for maximum efficiency and mechanised harvests, most of our olives were originally planted in wide rows with generous gaps in between, something I regularly thank the forebears for. It allows each tree to mature into a magnificent, gnarled personality of its own and the grove itself – the space between the olives – hums with life.

The avenues are so wide I've often considered cultivating legumes between the trees, although that would mean fencing them off from wild boar, who constantly overturn the soil, providing a year-round gardening service. In recent years, though, I've put almost as much effort into multiplying the diversity of wildflowers and edible weeds as I have into maintaining the olives. Many in the area do the opposite: I see groves rising from bare, stony earth, extirpated of all other plant matter in the lingering belief that all vegetative competition must be eliminated for the sake of the olive. I usually don't even flail the dead vegetation until close to harvest time, trapping as much vital moisture as possible in the undergrowth during the hottest months of the year.

But in truth, these weeds and wildflowers are half of Italy's food story. Soon after the first wild asparagus (*Asparagus acutifolius*) appears, often growing at the very base of the olives, the rest of the grove begins to awaken. There are dandelions (*Taraxacum officinale*), still gathered for spring salads or boiled and dressed with lemon and oil. The beautiful bright purple tassel hyacinth (*Muscari comosum*) hides its edible bulb beneath the surface and is dug up in southern Italy to make *lampascioni*. There are also chicory (*Cichorium intybus*), wild fennel (*Foeniculum vulgare*) and, in the shadier spots, good King Henry (*Blitum bonus-henricus*), an ancient spinach relative. In some years, I've even spotted mallow (*Malva sylvestris*) and the occasional wild garlic (*Allium vineale*), though that's rarer in our dry hills.

I'm still relatively new to foraging, and I'm sure I'm missing even better things to eat. But professionals have told me that much of what grows spontaneously among our olive trees is not only edible but among the most nutrient-dense plants on Earth. The olive trees stand like proud parents above a living understory, a carpet of abundance, most of which is criminally overlooked. So much of this wild growth is either edible, medicinal or valuable to pollinators.

In a land that hasn't always been as plentiful as it is today, rural Italians once knew these plants as intimately as their cultivated crops. For centuries, wild greens, roots and shoots were lifesavers, gathered in years when harvests failed or were plundered by marauders. Hunger, particularly in southern Italy, frequently stalked the land. Over generations, those flavours of the wild – bitter, resinous, mineral, astringent – became woven into the national palate. Italians still prize *amaro*, bitterness, and depth over sweetness: tastes shaped as much by resilience and necessity as by abundance.

This inheritance reaches back to the monasteries, where herbal knowledge once bridged medicine, devotion and survival. The monks' apothecaries – those same gardens that distilled remedies and tinctures – became the cradle of Italy's enduring taste for the bitter and the botanical. The tradition of *amari* – bitters – from Fernet-Branca to the wild herb liqueurs of Alpine and Apennine valleys, is a distilled expression of that memory: a spiritual and sensory lineage linking the dinner table to the hillside, the forest and the ancient instinct to taste the earth as it really is.[5]

Beyond the olive groves and at the margins of our territory, I've found nearly all the major native fruit, nut and berry varieties that would have been known and foraged across early human history in Italy. In between the cork and holm oaks we have the wonderful *corbezzolo* (*Arbutus unedo*), or strawberry trees, whose bright red, if mostly tasteless, fruits have been collected since antiquity. The fruits have a tendency to ferment while still on the branch and the overripe ones, when eaten raw, can induce a light inebriation, hence the Latin name *unedo* – usually attributed to Pliny the Elder, who glossed it as *unum tantum edo*, 'only one I eat'. In Sardinia and parts of Tuscany one can obtain *corbezzolo* honey, considered one of the most precious of all.

Figs (*Ficus carica*) are one of the few fruits still cultivated today that can truly be called native to Italy. A plant of ancient lineage and deeply embedded in the Mediterranean landscape, it grows wild on rocky terraces, stone walls and parched hillsides. I've often spotted its unmistakable leaves sprouting miraculously from cracks high in the masonry of ruined churches and farm buildings.

Figs have been selectively cultivated since at least the Bronze Age, probably even earlier, and are among humanity's oldest and most enduring allies. They were one of the first plants ever domesticated, propagated not from seed but by cuttings, a process that bound human hands directly into their reproduction. Each fruit is also the product of a delicate symbiosis: pollinated by a single tiny wasp that can survive only within its flowers. Across the Mediterranean, figs fed shepherds, soldiers and monks alike, providing sweet, durable energy where few other crops would grow.

Archaeobotanical remains from southern Italy and along the Tyrrhenian coast show early fig cultivation alongside olive and grape. By the time of the Etruscans and Romans, the fig had become not only a staple crop but a symbol of fertility, abundance and domestic well-being. Pliny the Elder lists dozens of varieties grown in different regions of Italy and fig trees were often planted at the heart of Roman gardens, as much for shade and beauty as for sustenance. If you've ever wandered the top floor of the Palazzo Massimo in Rome, surrounded by the exquisite frescoes of the Villa of Livia, you can almost picture the scene. Our own wild fig trees, spread across the farm, feed us almost continuously throughout August.

Besides the figs, I've found myrtle bushes (*Myrtus communis*), hawthorn (*Crataegus monogyna*), juniper (*Juniperus communis*), wild vines (*Vitis vinifera* subsp. *sylvestris*), wild vetches (*Lathyrus*) and peas (*Pisum*), buckthorn (*Rhamnus alaternus*) and sloes (*Prunus spinosa*) growing intermittently in and beyond the farm. There is evidence of nearly all of these being foraged in the last ten thousand years.[6] In a few quiet corners, there are also sorbs, or service trees (*Sorbus domestica*), which were once among the most common food trees in Italy. Their medlar-like fruits were traditionally stored in the dark and eaten slowly over the winter months, the astringency softening with time. Today, they are all but forgotten.

I've even come across a Cornelian cherry (*Cornus mas*), whose dense, fine-grained wood was highly prized in antiquity for tool handles, spears and – according to legend – even the frame of the Trojan Horse.[7] Here in the Maremma, the local cowboys, or *butteri*, carry a curved stick – an *uncino* – traditionally made from cornel wood. The fruit of the *corniolo*, tart and tannin-rich, was once used medicinally as an astringent and a remedy for dysentery. Closer to the streams, clumps of elderberry (*Sambucus nigra*) thrive, and deeper into the woods I've found stands of wild hazel (*Corylus avellana*), one of Italy's only truly native nuts.

Hazelnuts were once a minor but precious forest food, picked from shaded groves. Their Latin name refers back to Avellana, near Naples, which has been synonymous with hazelnut cultivation since antiquity. Today, they cover roughly 90,000 hectares of Italian farmland, mostly planted in monoculture. Too often, many of these intensively managed orchards rely on irrigation and agrochemicals in vast quantities, accelerating soil degradation and sending nitrogen and pesticide runoff into waterways. Many have also rendered themselves highly vulnerable to an impressive array of pests and diseases, most notoriously the brown marmorated stink bug (*Halyomorpha halys*), whose appetite for hazelnuts themselves rivals our own for hazelnut spreads. The predictable response, of course, is to reach for more pesticides.

The volcanic Lake Vico in northern Lazio is a grim case in point. With more than 3,100 hectares under hazelnut cultivation in its catchment, the lake has suffered severe eutrophication and recurring toxic algal blooms. In a series of landmark rulings, Italy's highest administrative court found that regional authorities had failed to protect this Natura 2000 site and drinking-water source from agricultural pollution, and ordered urgent corrective measures. Once pristine, the lake's waters are now frequently unsafe for use – a warning of what unchecked intensification can bring.

Yet it doesn't have to be this way. Hazelnuts can be grown organically, even at scale, and without irrigation in many Italian regions. Organic certification and sourcing standards could be implemented that prevent the kind of ecological degradation now unfolding, helping to preserve landscapes and sustain a crop that is as ancient and native to Italy as the forests it once grew beneath.

Lightning Nodules

Beyond the cultivated crops and modern plantations, however, some of Italy's most prized foods still come not from fields or orchards, but from the stealthy, unpredictable generosity of the wild. Undoubtedly, the wild fruiting bodies that inspire the most seductive excitement, jealous secrecy and eye-watering prices are the funghi. After the first rains in late summer, the land around us is visited once again by a cast of familiar local figures, returning to their favoured, half-secret hunting grounds. They come for the *finferli* (chanterelles), porcini, *ovoli* (Caesar's mushroom), *trombette dei morti* (horns of plenty) and a host of other varieties, depending on their knowledge. Some knowledge, however, is essential. Among the most delicious mushrooms, I have found nearly all of the most lethal: death caps, panther caps, destroying angels and deadly webcaps, often nestled alarmingly close to prized edibles.[8]

Italy is thought to host at least 15,000 species of fungi, an unseen kingdom of astonishing diversity. Of these, around 1,200 are considered edible; perhaps 100 to 150 are widely gathered and eaten, and only 40 or so are treated with particular reverence. Roughly 500 species are toxic, and at least 40 are lethally poisonous. The odds are therefore stacked against the unwary.

There is one type of fungus that, unfortunately, doesn't seem to encroach on our territory, and which for us will remain what it is to most people: an expensive treat – the truffle. Italy is the undisputed land of truffles, thanks to the extraordinary variety of its soils, climates and tree associations. Nowhere else do limestone, clay and forest combine in such intricate patchworks, from the oak and hazel groves of Piedmont to the beech and holm oak woods of Umbria, Tuscany and Abruzzo. The country's long history of mixed woodland management, neither entirely wild nor fully cultivated, has also proved ideal for these elusive underground fungi, which rely on symbiosis with tree roots and an undisturbed, well-aerated soil.

Because truffles fruit underground, they can't rely on wind for spore dispersal. Instead, they evolved to produce their signature pungent aroma – earthy, musky, sometimes garlicky – to seduce forest mammals such as wild boar, foxes, squirrels and deer. Female pigs were once favoured for truffle-hunting because the scent mimics androstenone, a pheromone found in boar saliva. They dug, ate and spread the spores – a subterranean partnership that has endured for millions of years. Today, pigs survive mainly as tourist attractions, while trained dogs, especially the Lagotto Romagnolo, Italy's native truffle-hunting breed, do the real work. They're not entirely innocent, of course: many can't resist eating their finds, but unlike pigs, they're small enough that a quick-handed owner can usually intervene before the prize disappears. Together they sustain a national truffle industry worth some €200 million a year.

The peninsula harbours more than twenty-five native species of the truffle genus *Tuber*, including all of the most prized varieties in the world. Some, like the fiercely guarded white truffle of Alba (*Tuber magnatum*), among the world's most expensive foods, grow only in the moist forests of Piedmont, Tuscany, Le Marche, Umbria and Molise on calcareous soils. Others, like the black truffle (*Tuber melanosporum*), are more adaptable, widespread and even cultivated in parts of central Italy. The bianchetto (*Tuber borchii*) thrives in coastal pine woods; the *scorzone* (*Tuber aestivum*) fruits in summer and again in autumn, its aroma milder, less expensive but still unmistakable.

This hyper-regionality – each truffle bound to a precise marriage of tree, soil, moisture and climate – mirrors the very structure of Italy itself: a mosaic of microclimates and terrains, where limestone meets lava and where ancient forests and Mediterranean scrub coexist within a few kilometres. The truffle's secret life is therefore also the story of Italy's landscape – its restless geology translated into aroma and flavour, distilled from underground.

The Romans, following Pliny the Elder, believed truffles were born of lightning strikes – little heavenly nodules springing from the irate flashes of Jupiter's thunderbolts. To the medieval mind, they were the work of the devil – dark fruitings of a diabolical world, fit only for witches and heretics. For centuries they were spurned, feared and forbidden. Then came their slow redemption: revived during the Renaissance, elevated in royal kitchens, and by the nineteenth and twentieth centuries, crowned at the very top of gastronomy's food chain.

I often wonder why truffles, even the ubiquitous summer variety, seem absent from the land immediately around our farm. Perhaps our soils simply lack enough limestone, or are a little too acidic. Perhaps we lack the right mix of tree species or our summers are simply too hot and dry – the Maremma is among the most arid parts of Italy and we don't get quite enough autumn rain to coax them from the soil. Maybe they were once here, but the tradition of looking has faded. Or perhaps they are here, quietly thriving, and the locals are simply staying tight-lipped. More than once, I've been tempted to invite a visiting truffle hunter and his dog to the farm, just to settle the matter once and for all.

Arrivals

Wild and foraged foods, however, tell only a fraction of Italy's culinary story. The deeper truth is that Italy today possesses one of the highest concentrations of agricultural biodiversity on Earth, especially when measured relative to its size. Thousands of years of cultivation, migration, conquest and exchange have yielded an extraordinary abundance of cultivated fruits, vegetables, grains and legumes – a richness of local varieties and landraces that has few parallels anywhere in the world.[9]

This abundance has deep roots. Italy sits at the crossroads of three biogeographical regions – Alpine, Mediterranean and Balkan – and its topography creates an almost endless range of microclimates. From snowbound valleys to volcanic coasts, crops that would perish elsewhere find niches here. Every migration and empire added new seeds: Phoenicians and Greeks expanded the cultivation of olives and vines; Arab traders and farmers introduced citrus, sugarcane and aubergines; the Romans diffused grains and pulses across the peninsula; and after Columbus, maize, tomatoes, potatoes and peppers remade the country's diet. Layered upon this were centuries of isolation, mountain villages, islands and hill towns, where farmers selected and saved their own seed, giving rise to landraces – traditional crop varieties selected and saved locally over generations – perfectly attuned to their soils and seasons.

According to Italy's National Register of Varieties and the Slow Food Foundation's Ark of Taste, the peninsula harbours thousands of distinct traditional crop varieties and landraces, many of them grown nowhere else on Earth – a concentration of agrobiodiversity that exceeds that of any other European country and is widely regarded as among the highest in the world. By some estimates, well over 7,000 traditional crop varieties and landraces are recognised nationally, a figure compiled from overlapping national and regional catalogues.

Even within a single category, the variety is staggering: more than 1,200 grape varieties; around 500 olives; over 300 tomato landraces; over 200 wheats; and hundreds of legumes, from Castelluccio lentils to Lamon beans and the chickpeas of Navelli. There are scores of rice varieties, dozens of citrus and artichoke cultivars, many tens of chestnuts, and well over a thousand apples, from alpine valleys to the Apennines. This genetic abundance is deeply entwined with Italy's cultural diversity, shaped by generations of farmers, monks, migrants and breeders who tuned their crops to the subtleties of place. In doing so, they made place inseparable from taste – and turned geography itself into gastronomy.

From the fifteenth century onward, wealthy landowners often celebrated the abundance of fruits and vegetables grown on their estates in huge paintings and frescoes commissioned from the great artists of the day. In a quiet corner of Umbria, a remarkable woman named Isabella Dalla Ragione is using those same paintings and frescoes to help rescue Italy's forgotten fruit trees from extinction. For decades she has traced clues in Renaissance paintings and monastic archives, following the trail of gnarled remnants in abandoned orchards to recover the last survivors of Italy's rarest cultivars. Her living museum, high above the tobacco and hazelnut monocultures of the Tiber Valley, now shelters more than 600 trees and over 150 varieties of fruit, each a genetic time capsule rooted in a particular place and moment of Italy's past.

Until the twentieth century, nearly every region of Italy nurtured its own distinct apples, pears, cherries and plums, each shaped by local soils, climates and uses. By the early 1900s, Italy still boasted more than a thousand varieties of pear alone. Then, industrial agriculture and supermarket uniformity swept much of that diversity away. Today, just four apple and pear varieties account for more than 70 per cent of national production. With each lost variety, Italy loses not only flavour but resilience: the adaptive traits that once allowed crops to weather drought, frost or blight. At a time when climate change is reshaping agriculture across the planet, such genetic diversity may prove one of the most valuable resources humanity possesses.

Nowhere are the costs of this narrowing clearer than in South Tyrol's Val Venosta, Europe's largest continuous apple-growing region. There, researchers have found residues of twenty-seven different pesticides in the surrounding mountain environment, even in children's playgrounds. In these single-variety orchards, diseases like apple scab and fire blight spread unchecked, sustained only by relentless chemical regimes. Agricultural runoff has become to Italy's landscapes what industrial pollution once was to its rivers.

Against this backdrop, the quiet work of people like Dalla Ragione takes on the weight of resistance and renewal. By preserving Italy's ancestral cultivars – not only fruit trees but grains, legumes and vegetables – she and others like her are safeguarding the country's genetic future. In rediscovering the plants that once sustained its past, they are helping Italy reclaim the most important ingredient of all: resilience.

From the olives of Puglia to the hazels of Lazio and the apples of Trentino, the pattern repeats: large monocultures with low genetic diversity encounter a new pathogen to which they have no defence, resulting in near-total collapse unless drowned in pesticides. Yet in the case of olives, Italy's deep reservoir of genetic diversity, some five hundred recognised cultivars, offers a glimmer of hope. A few varieties, most notably Leccino, one of the three I grow myself, are showing signs of resistance to *Xylella*. It's a small comfort to know that perhaps a third of my trees may survive when – if – the disease reaches Tuscany. But the lesson is clear: resilience is born of diversity. In a world where trade, travel and climate now connect everything everywhere all at once, our best defence against the tide of invasive diseases and pests is to preserve the widest possible pool of life.

Near and Far

When we bought our small farm in Maremma, beyond creating our own little Italian refuge, I had one quiet ambition: to live off the land as much as possible. What I couldn't grow myself, I would try to source locally. Italy is a good place for such an ambition, not just for eating Italian, but for eating locally. Wherever you go, it's nearly always possible to prioritise regional markets, small grocers and direct sales from farmers. Even the supermarkets, which still coexist with strong traditions of markets, offer clear sourcing and seasonal produce from nearby fields.

Within weeks of arriving, I had struck a deal with a local farmer: a supply of free-range chickens, half a pig at New Year, the occasional rabbit and sometimes beef or lamb when available. The same farmer, also head of the local hunting team, noted our location and made sure I also unfailingly receive a haunch of wild boar every December. Not long after, I discovered one of Italy's oldest organic farms, just ten minutes away, growing some of the best vegetables I've ever tasted. Their surplus is bottled, fermented and preserved into a galaxy of store-cupboard jars that nourish us throughout the year. Nearby, another farm was specialised in organic dairy, putting out some of the most delicious pecorino and yoghurt I know.

My greatest aversion has always been to buying fish, most of it caught far away, often by destructive bottom-trawling fleets with little regard for bycatch, seasonality or ecological limits. But then I heard about Paolo il Pescatore. A legend along the Tuscan coast, Paolo has resisted the industrial fishing fleet since the 1980s, even creating an underwater sculpture park off the coast of Talamone designed to make trawling physically impossible.[10] He's become a local hero of mine, not least for his valiant, single-handed resistance to one of the most environmentally destructive industries on the planet, which is steadily reducing large parts of our oceans to marine deserts. But also because he still fishes from a small boat, which he welcomes visitors aboard; with restraint, seasonal awareness and deep knowledge of the sea. His is practically the only fish I buy and eat anywhere in Italy.[11]

Where we are in southern Tuscany isn't even particularly exceptional. I've lived for four years in Milan, and spent both long and short stretches in Venice, Florence and Rome. I've stood wide-eyed and salivating in food markets in Bologna, Naples, Palermo, Ancona and Turin. Without a doubt, one of the greatest privileges of living in Italy is the ability to source, almost effortlessly, some of the best-tasting, most seasonal and most locally rooted produce anywhere in the world. And you can do this to varying degrees in every single village, town and city up and down the country. Whenever I've had an infuriating experience in some grey administrative office, opened another incomprehensible tax bill, or nearly been run over crossing the road, I find immediate comfort and clarity at the food market. It's where Italy makes sense again.

Ultimately, though, I've failed to live entirely off what Italy itself produces. Coffee is so deeply entwined with daily life, so seamlessly stitched into routine, that it has proven impossible to renounce, despite the fact that not a single bean is grown on Italian soil. I've tried the most common substitute offered in bars: *caffè d'orzo*. A staple during wartime shortages, it's a roasted barley brew with none of coffee's caffeine but all of its ritual, served in the same small white cups, sipped slowly and taken semi-seriously. I even once experimented with a homemade version using roasted carob pods, gathered from the squat, ancient trees that still grow wild in the far south.

The source of that *caffè d'orzo*, barley, is one of Italy's oldest and most venerable crops. It likely arrived around eight thousand years ago with the first wave of Neolithic farmers from Anatolia and the Near East, who likely followed both coastal and inland routes along the Adriatic, bringing their seeds, tools and domesticated animals. Barley quickly adapted to Italy's varied environments, thriving everywhere from the alluvial plains of the Po to the dry hills of the south. It remains a significant cereal crop today, not least as the base of that unsatisfying coffee substitute.

Barley probably arrived alongside a small company of other pioneering crops: lentils, peas, flax, millet and even poppies. Together, they formed the first Mediterranean toolkit of cultivation: plants hardy enough to cope with thin soils, summer drought and the peninsula's restless patchwork of microclimates. Their success laid the foundations of Italy's earliest agrarian settlements, whose rhythms of sowing and harvesting would endure for millennia.

Most transformative of all, though, was the grain that would come to define the Italian table: wheat. Of all the crops ever grown in the country, none has left a deeper imprint on its landscape or its identity. From the terraced fields of Sicily to the plains of Emilia-Romagna, wheat has shaped everything from diet and economy to language and ritual. It is the backbone of Italian nutrition, and the staff on which the whole story of pasta, pizza and *pane* still rests.

The most ancient wheat variety to arrive in Italy was einkorn: *farro piccolo* (*Triticum monococcum*), one of the first cereals ever domesticated in the Fertile Crescent. It likely arrived around the same time as barley, around eight thousand years ago with migrating Neolithic farmers from the Near East. This was followed by emmer: *farro medio* (*Triticum dicoccum*) and later spelt: *farro grande* (*Triticum spelta*), probably introduced from northern Europe. Both the ancient Romans and modern Italians collectively refer to these three primitive grains as *farro* (*far* in Latin). However, if you ever have the pleasure of eating a nutty *insalata di farro*, *pasta di farro* or indeed a *farotto* (farro risotto), it will almost always be made from emmer wheat – *farro medio* – which is still grown today in substantial quantities across Tuscany, Umbria, Le Marche and Abruzzo.[12]

In terms of a nation's nutrition, emmer wheat may be the most historically significant crop ever grown in Italy. For roughly five thousand years, from around 6000 BCE until the early Roman era, it was the primary cereal cultivated across the peninsula. Hardy and adaptable, emmer thrived in nearly all of Italy's diverse soils and climates. Its story, however, begins far from Italy. Emmer has deep evolutionary roots in the Fertile Crescent, where it emerged through a natural hybridisation between early wheat ancestors and a wild goatgrass (*Aegilops*) species long before humans began farming. Around ten thousand years ago, early agriculturalists in the Near East domesticated this wild emmer, selecting for grains that held to the stalk and yielded better harvests. The result became the backbone of prehistoric agriculture, eventually taking root in Italy's fields and rituals.

The oldest known evidence of wheat cultivation in Italy comes from Vhò di Piadena, near Cremona in Lombardy.[13] Here, in the fertile floodplains of the Po Valley, formed by millennia of Alpine sediment and enriched by regular flooding, archaeologists uncovered charred grains of emmer wheat alongside barley, einkorn, lentils and a curious two-headed female figurine who probably presided over the land's fertility.[14] Dated to around 5600 BCE, these finds mark the earliest signs of Neolithic Italians sowing, harvesting and milling wheat: the beginnings of an agricultural pact between humans and landscape.

That pact would define Italy's identity for millennia to come. Wheat, alongside the vine and the olive, became the agricultural trinity that bound Italians to their soil – sustenance, ritual and trade intertwined. From the Bronze Age onward, the rhythms of ploughing, threshing and baking shaped not only the countryside but its very culture, leaving traces in language, festivals and faith. It would take nearly seven millennia – and the slow refinements of the medieval world – before wheat found its most iconic Italian expression: pasta.

However, in those earliest years, long before pasta, the most likely use of emmer and other grains was as a simple porridge: ground grain boiled with water or perhaps milk, if it was available. The Romans would later give this gruel the name *puls*, a dish that, in one form or another, may have sustained more people in Italy longer than any other. From the first stone querns to the hearths of Etruscan farmsteads, this humble preparation carried the weight of survival, becoming the unglamorous ancestor of Roman cuisine. That may sound like an extravagant claim, yet *puls* never really disappeared. It simply evolved. Replace emmer or barley with millet, spelt or later maize, and you have polenta – the same elemental formula of grain and water. Across millennia, through poverty and plenty alike, this simple, adaptable dish has arguably fed more Italians, for longer, than any other food in history.

Although emmer was likely used to make simple unleavened flatbreads, it wasn't until the late Republic, around the second century BCE, that leavened bread-making began to flourish. By this point, Rome's conquests abroad had allowed it to begin importing new varieties of wheat, including the ancestors of the two principal types still grown today: bread wheat (*grano tenero*) and durum wheat (*grano duro*). Emmer, which had reliably fed Italians for millennia, gradually slipped into secondary status, retreating to the mountains and margins as the demands of urban milling, imperial logistics and a rising appetite for soft, leavened bread elevated these free-threshing wheats to dominance.

For more than two thousand years, durum and bread wheat have shaped the Italian landscape. Bread wheat dominated in the cooler, wetter soils of the north and central regions, while the hardier, drought-tolerant durum wheat blanketed much of the south, particularly Puglia, Campania and Sicily, where it still defines the country's pasta culture. Up until the twentieth century, Italy's wheat fields were sown with a tapestry of ancient landraces, each adapted to local soils and climates. In Sicily, hardy strains like Russello and Timilia flourished without irrigation. In the highlands of Abruzzo and Tuscany, peasants grew Solina and Gentil Rosso: wheats with red grains and delicate flavour, often stone-milled and baked into dense, rustic mountain breads. In the north, especially the Po Valley, varieties such as Frassineto were cultivated, known for their adaptability and suitability for soft bread flours. It's thought there are still about 250 varieties of wheat, both ancient and modern, grown in Italy today.

The main problem in Italy has always been one of supply and demand. Until the twentieth century, wheat yields of less than 1 tonne per hectare, while rising elsewhere in industrial Europe, remained virtually unchanged since Roman times. The Romans of course, at the height of the empire, depended on vast imports of grain from Sicily, Sardinia, Egypt and

North Africa to feed their cities. After the empire's collapse and the sharp decline in population, grain imports were no longer necessary and subsistence agriculture met local needs. But from the High Middle Ages onward, growing urban centres like Venice, Genoa and Pisa once again outstripped their agricultural base and turned to the sea, importing grain on special fleets from the Dalmatian coast, Sicily and beyond. By the eighteenth century, nearly every major Italian city relied on imported grain and hunger frequently blighted the land.

Wheat's importance was absolute. More than any other crop, it sat at the centre of the Italian diet, the economy and even social order. When harvests failed, there were no easy substitutes: bread was the daily foundation and the loss of wheat meant immediate scarcity. Poor yields could trigger soaring prices, food riots and famine. In a landscape that had learned to live by its harvest, the fragility of a single crop could ripple through every level of society.

At the beginning of the twentieth century, the government of the newly unified Kingdom of Italy set out to tackle the country's long-standing wheat problems: low yields, disease vulnerability and dependence on imports. One of the major agronomic challenges since ancient times had been the extraordinary height of traditional wheat varieties. These older wheats often grew between about 90 and 180 centimetres tall – significantly taller than modern cultivars – making them prone to lodging: collapsing under their own weight, especially in wind or rain, frequently ruining entire harvests just before maturity.

Long before the Green Revolution of the 1950s, a humble agronomist from Le Marche, Nazareno Strampelli, began hybridising wheat varieties, first near Rieti and later in experimental fields around Foggia in Puglia. Working under both liberal and fascist governments, Strampelli developed around

65 wheat varieties, and by the 1930s his strains accounted for nearly 70 per cent of all wheat grown in Italy.

His wheats spread well beyond the borders. Several varieties made their way to Mexico, where they became part of the genetic foundation for Norman Borlaug's wheats in the 1960s as part of the Green Revolution. Two of Strampelli's creations became especially renowned. First, the soft bread wheat Ardito, released in 1913, and, second, his most iconic variety, the durum wheat Senatore Cappelli, introduced in 1915. It became the cornerstone of Italy's pasta industry. With yields increasing by over 50 per cent, Strampelli's work made Italy, for much of the first half of the twentieth century, nearly self-sufficient in grain. Although long since replaced by modern high-yield varieties, his creations laid the foundation for Italy's twentieth-century wheat revolution – one whose success helped transform a country that had regularly failed to feed itself into a nation that now helps feed the world.[15]

It won't surprise anyone that Italy is the world's largest pasta producer. Few foods illustrate a country's ability to transform simple ingredients into cultural identity quite as completely as pasta. Today, Italy produces more of it than any other country on Earth, turning out over 4 million tonnes a year and supplying much of Europe's demand. Italians themselves remain the world's most devoted consumers, eating roughly 26 kilograms per person annually.

Wheat, however, tells a more complicated story. Of Italy's over 12 million hectares of farmland, a little under one-fifth is planted with soft and durum wheat. Domestic harvests, although substantial, are never enough: mills and pasta factories together grind more than twice what Italian fields produce. As a result, Italy has become one of the world's major grain importers, drawing in shiploads of durum from

Canada and soft wheat from France, North America and Ukraine. Paradoxically, some of that imported grain leaves again as premium Italian flour – a subtle reassurance to overseas pizza lovers that their flour, at least, passed through Italian hands, if not Italian fields, on its way to the oven.

Home Grown

In an age when most nations depend precariously on global supply chains, Italy remains something of an anomaly. Despite its modern crises and shrinking rural workforce, the country still produces a large share of its food at home. Deep-rooted regional systems, diverse crop varieties and a cultural devotion to seasonal eating make Italy far less vulnerable to sudden trade shocks than its modest size might suggest. In a world grown uncertain, that resilience, born of diversity, may yet prove to be one of Italy's greatest natural resources.

I take real joy in the simple rhythm of the market: nearly every month brings something new to the stalls. I've walked chefs through humble farmers' markets and watched them well up at the sheer variety, quality and affordability of what's on offer. In January, the spiky purple artichokes from Liguria and Sardinia arrive; in December, it's the crisp, bitter *radicchio rosso* from the Veneto; February brings the blood oranges from Etna – an essential jolt of vitamin C when you need it most. November is for chestnuts, *castagne*, from Monte Amiata. In spring, mountains of fava beans and peas pile high; in summer, it's the wrinkled Cuore di Vesuvio and San Marzano tomatoes, striped zucchini with their flowers, flattened red onions from Breme in Lombardy or oblong pink ones from Tropea in Calabria, as well as the abundance of stone fruit that bring the colour.

Much of this strength lies in the persistence of the small-holding model. For all the pressures of industrialisation and rural depopulation, Italian farms remain largely family-run, diversified and embedded in their landscapes and communities. Inheritance laws introduced under Napoleon, which required land to be divided equally among heirs, created a patchwork of small plots. Over time, that mosaic helped safeguard agricultural biodiversity, regional food cultures and a sense of stewardship that tends to fade where land is consolidated into few hands. Fragmented though it may look on a map, this pattern of ownership has ensured that the land is tended, known and passed on – and that it continues to feed people in ways that connect them to place.

Most heartening of all is a quiet generational shift now underway. After decades of exodus, thousands of young Italians are returning to the countryside their grandparents abandoned. More than fifty-five thousand people under the age of thirty-five now work in agriculture – the only sector in Italy to show sustained growth in youth-led enterprises. Many are leaving urban careers to pursue small-scale projects that blend regenerative farming, agritourism and direct-to-consumer ventures, driven by a search for meaning, autonomy and a healthier rhythm of life. Seventeen new youth-run farms are founded each day: a trickle, perhaps, but one that carries the unmistakable scent of renewal.

For a country long defined by its relationship to the land, these are modest but hopeful signs – possibly the first green shoots of a return to abundance.

EPILOGUE

The Nature of Things

> If these pages have the luck, unfortunately too rare, to leave the walls of city schools, to spread in the countryside, in the heart of the Alps, in the mountains of the Apennines, at the foot of Vesuvius and Etna, they will teach the inhabitants of those districts to appreciate themselves a little better and the beauties and favours of every kind of which nature, minister of God, was not stingy to the various provinces of Italy.
>
> **Antonio Stoppani,** *Il Bel Paese*

It was barely ten rungs up the rusting ladder that I began to question the wisdom of climbing the abandoned cable car pylon. I wasn't even a quarter of the way to the top, and already every instinct was telling me to turn back. A bat shot out from a recess, its silken wing brushing across my face. I flinched, clinging on with one arm while nervously fumbling with my phone in the other, trying to light the concrete shaft. I could only hope the grey hollow tower I was climbing was still sound. The owner of the bed and breakfast had promised it offered the best view of the lakes, recommending it with such casual flippancy that retreat now felt like cowardice – despite every alarm bell in my body insisting otherwise. It looked like it had stood since the 1960s; if it hadn't fallen yet, I reasoned, why would it fall today?

The pylon I was climbing belonged to the old cable car that once carried passengers from the shores of the Monticchio lakes to the summit of Monte Vulture, a dormant volcano in Basilicata, near the borders of Campania and Puglia. Built in the 1960s during the post-war tourist boom, it was abandoned barely two decades later, when visitors began drifting elsewhere – to the coast or abroad – leaving this oasis to decline economically but improve ecologically, reverting to a wilder, more natural state. The B&B owner had been right: the view was spectacular.

Of all the places I have visited in Italy, Monte Vulture seems to distil the country's natural history more completely than any other. An Ice Age volcano, it began erupting about one million years ago, not long after the surrounding land had risen from the sea. Over time, it passed through many moods, from explosive Plinian outbursts to gentler, more effusive phases, before falling silent around forty thousand years ago. Defying the usual pattern of Italian volcanism, Vulture stands apart to the east of the Apennines, an isolated outlier of the back-arc volcanoes strung down the Tyrrhenian coast. Beneath its slopes lie sediments of a Pliocene sea, not so long ago in geological terms, that still yield the bones of whales which once swam here.

Within its humid, forested slopes, a rare microclimate shelters Europe's only surviving owl moth (*Brahmaea europaea*), perhaps Italy's most extraordinary endemic: a miraculous Ice Age survivor whose closest kin now mostly live in faraway Asian forests. In the surrounding lowlands, Palaeolithic hunters once tracked mammoth, straight-tusked elephant, hippopotamus and giant deer. By the Neolithic, people had settled on its flanks, leaving tools, pottery and ochre-painted figures of deer and humans at the nearby Tuppo dei Sassi rock shelter. In Roman times, Vulture's immense beech, chestnut and oak forests were cut for timber, its slopes cleared for crops, only for nature to reclaim much of the land in the centuries that followed.

By the early Middle Ages, Byzantine hermits had carved cave dwellings above the Monticchio lakes, transforming the site into a place of sanctuary. Their community grew into a monastery, today housing one of Italy's most beautiful natural history museums. All the while, the volcano's rich soils have nourished vines, producing one of the country's great wines, Aglianico del Vulture, the so-called Barolo of the south. Beyond the confines of this volcanic oasis, endless fields of durum wheat stretch across the horizon.

That night, I lay in the silence of my bed and breakfast, listening to the howls of wolves and the hurried hooves of wild boar and roe deer on the cracked asphalt outside my window. My thoughts drifted back to the afternoon, when Renato Spicciarelli, Professor of Entomology at the University of Basilicata and the foremost authority on the owl moth, had guided me through the remarkable natural history museum above the lake, a place he and his colleagues had created with evident devotion, despite limited resources. I couldn't help but conclude that monasteries, when no longer sustained by monks, make excellent natural history museums.

Places like the Certosa di Calci near Pisa, a former Carthusian monastery that now houses Pisa University's exquisite Natural History Museum, or Alba's excellent archaeological and natural history museum, set within the former convent of Santa Maria Maddalena, show how calm, contemplative spaces seem innately suited to preserving and presenting the wonders of the natural world. In Rome, however, despite repeated appeals from members of the Accademia dei Lincei, there is still no national natural history museum – a baffling omission in a country that arguably gave birth to the discipline. Vast collections of Italy's palaeontological and natural heritage remain hidden from public view, scattered in basement archives, at a time when a broader appreciation of our natural world has never been more urgent. And in Rome, after all, there is no shortage of half-empty monasteries waiting for a new purpose.*

Yet museums alone are not enough: collections and specimens may preserve the past, but they cannot by themselves secure the future. For that, we need a population that is literate in the natural world.

It has become glaringly obvious to me that, alongside the core subjects, basic natural history, earth sciences and the story of the planet should be taught formally – at the very least in all primary schools. As well as being tested on times tables, spelling and grammar, every child should be fluent in the flora and fauna of their own surroundings by the age of ten. If we are to raise a population that cares about

---

* Italy is blessed with dozens of excellent regional natural history museums in nearly every major city, including the magnificent Renzo Piano-designed MUSE in Trento. Whether a new one in Rome could add much is debatable – but it could serve as a unifying hub for all of them, mounting blockbuster natural history exhibitions that draw on collections and expertise from across the country, and galvanising the national scientific community as a whole.

the environment and is to help reverse the damages to the planet's health wrought by previous generations, then a rudimentary grounding in natural history is essential. I have never encountered a child without an innate curiosity for the natural world. A few hours each week dedicated to earth science, both in and outside the classroom, could nurture that spark into knowledge, empathy and responsibility. It seems to me one of the most urgent priorities of our time.

I rolled over on my bed, the sounds of the night filling my room, thoughts and ideas swirling in my head. I began retracing my steps from earlier that day through the whitewashed corridors of the Abbey of San Michele Arcangelo, which houses the Museo di Storia Naturale del Vulture. Beyond the displays and stories, what stayed with me most was a poster at the entrance, proudly announcing the museum, but emblazoned at the top with the words, 'Appennino Parco d'Europa' – the Apennine Park of Europe.

I had asked Renato about it, assuming it was just a long-forgotten pipedream, an ambitious project that had never made it beyond the planning stage. 'It was,' he admitted, 'but the idea was so well loved, especially in the parks and reserves of the south, that we kept the dream alive longer than most.' The hope, he said, was that one day it might be revived, when there was more support and more money. A few days later, in the offices of the Pollino National Park in Calabria, I heard the same story: it had once been a bold dream, but it died for lack of energy to carry it forward.

First proposed in 1995 by the environmental association Legambiente, Appennino Parco d'Europa (APE) was conceived not as a single vast park, but as a coordinated network linking the national parks, regional reserves and cultural landscapes along the length of Italy's mountainous spine. Its aim was to treat the Apennines as one continuous ecological and cultural system – a living corridor of biodiversity, watersheds and rural heritage, rather than a scattering of isolated protected pockets. The Apennines are already one of Europe's most important faunal corridors, allowing wolves, bears, deer and countless other species to move between habitats in a way few other Western European landscapes can still support. However, the main thing preventing the Marsican brown bear from further expanding its range is better connectivity between the patchy protected areas.

Despite Italy's formal commitment to the EU target of protecting 30 per cent of land and sea by 2030, the expansion of its protected territories has proved anything but straightforward. Since the 1990s, several new national and regional parks have been created and the nature protection network Natura 2000's coverage has grown, yet the establishment of further national parks – genuine public conservation areas where, unlike in Britain, activities are strictly limited – has in most cases reached a political and administrative impasse.

These are notoriously complex entities to define and create. Three new national parks exist, albeit only on paper. The proposed Portofino National Park in Liguria, designed to expand the current regional park, has been stalled since 2021 amid wrangling between the Ministry of Environment, the region and local municipalities. The Costa Teatina National Park, intended to protect a 40-kilometre stretch of southern Abruzzo's Adriatic coastline, was first proposed in 2001 and even provisionally delimited in 2014, only to be halted by a Constitutional Court ruling reaffirming the state's sole authority to establish national parks. In Sicily, plans approved in 2007 for parks in the Egadi and Aeolian archipelagos and the Iblei mountains have remained frozen, with the regional government failing to act despite a 2009 court ruling upholding their legitimacy. As things stand, the tiny island of Pantelleria remains Sicily's only national park.

These three stalled cases reveal how politically and administratively fraught the creation of national parks can be, particularly in profitable coastal zones, arguably the very places where protection is most urgently needed. The Tuscan Archipelago National Park, encompassing seven volcanic islands, shows what is possible: it is witnessing one of the fastest recoveries of marine biodiversity anywhere in the Mediterranean. Just across the water, the Maremma Regional Park, also called the Uccellina, preserves a rare sweep of unspoiled coastline where deer, boar, wolves and ospreys still meet the sea. Yet despite Italy's identity as a country defined by its shores, less than 10 per cent of its protected areas lie along the coast. Put another way, out of roughly 8,000 kilometres of shoreline, only a few hundred are safeguarded within national or regional parks. In Sardinia, the vast Gulf of Orosei and Gennargentu National Park, decreed in 1998, has never been made operational due to intense local opposition, remaining a park in name only. The result is that much of Italy's coast remains exposed to unregulated development, mass tourism infrastructure and industrial activity, leaving fragile dunes, wetlands and marine habitats acutely vulnerable to irreversible loss.

By contrast, the process has proved less fraught in the remote, sparsely populated interior, where the recent creation of Matese National Park stands out as a rare victory. Upgraded from an existing regional park, it officially became Italy's twenty-fifth national park in April 2025 nearly four decades after first being proposed. The designation protects a mountainous massif straddling Molise and Campania, close to the fossil-rich strata of Pietraroja where Ciro, Italy's first dinosaur was unearthed, perhaps rekindling hopes of renewed palaeontological exploration in the area. Vast beech forests, karst caves, high-altitude lakes and habitats for wolves and bears make the Matese one of the most biologically rich corners of the central-southern Apennines. Its creation is a testament to persistence – and a reminder of the scale of the challenges that still confront conservation in Italy.[1]

The solution to creating new protected areas may lie in reviving the dream of the Apennine Park of Europe, but the ambition required would demand far greater coordination, incentives and flexibility than its predecessor ever achieved. The original vision already spanned nearly half of Italy's landmass, linking more than a thousand municipalities from Liguria to Calabria. A modern version could be more ambitious still: not simply a chain of parks and reserves, but an unbroken ecological artery running the length of the peninsula, weaving together rewilding zones, restored river systems, traditional farmlands, private reserves and historic villages. Such a project could align with the EU's Biodiversity Strategy for 2030 and the European Climate Law and be underpinned by emerging natural capital frameworks that treat healthy ecosystems as economic infrastructure on a par with roads, power stations, factories or even ski resorts. What it would need, above all, is collective enthusiasm – and no small amount of money.

Other regions have already shown what is possible. In Romania's Southern Carpathians, private philanthropists are helping to fund vast conservation landscapes, restoring large carnivores, native forests and river systems while supporting rural livelihoods. The ambition is to create a

'European Yellowstone' in the Făgăraș Mountains, with around 200,000 hectares earmarked for protection as a new national park. Though still aspirational, the vision of extending this model across the entire Carpathian range is now firmly on the horizon.

In South America, Kristine Tompkins, former CEO of Patagonia, and her late husband, Doug Tompkins, co-founder of The North Face and Esprit, acquired more than 800,000 hectares of threatened wilderness in Chile and Argentina, later donating the land to create or expand thirteen national parks. Their legacy safeguarded critical ecosystems, from Andean steppe to temperate rainforest, while also fostering local economies through ecotourism and regenerative agriculture.

Even in Italy, in the mountains of Biella, Piedmont, the textile entrepreneur Ermenegildo Zegna began in the 1930s to reforest the bare slopes above his hometown of Trivero (admittedly, slopes his company had helped to strip in the first place). Over the decades, his work grew into the Oasi Zegna, a 100-square-kilometre 'open-air reserve' that combines reforestation, nature conservation and sustainable tourism. It shelters the *Carabus olympiae*, an endemic jewel-green ground beetle whose survival is now actively managed through habitat restoration.* Today the Oasi is both a protected, restored landscape and an economic engine for the local community, showing how private investment, when tied to ecological restoration, can transform an entire mountain territory.

---

* Named in 1854 after Olimpia Sella, cousin of the celebrated mountaineer and statesman Quintino Sella, and, incidentally, ancestor of the Zegna family.

There is no reason the Apennines, taken as a whole, could not lead Europe in such an endeavour. This mountainous spine would instantly reassert itself as the continent's most important wildlife corridor and one of its most successful nature-positive economic zones. Some of the groundwork is already being laid through Rewilding Apennines, a public–private partnership working with local communities to expand corridors, restore habitats and build nature-based livelihoods, benefitting emblematic species such as the Marsican brown bear and the wolf. The potential to scale up is enormous. Instead of being mired in the slow, politically fraught process of creating new national parks, well-structured private or community initiatives could extend buffer zones around existing reserves, regenerate abandoned land and even establish entirely new protected areas. Such stewardship offers agility: weaving together habitat restoration, regenerative farming and ecotourism – a sector expanding rapidly worldwide. Yet to stand any chance of succeeding, it would require the combined forces of Italy's leading environmental and cultural organisations – Legambiente, WWF Italia, Lega Italiana Protezione Uccelli (LIPU, the Italian League for Bird Protection), Fondo per l'Ambiente Italiano's (FAI, National Trust for Italy), Slow Food – alongside the countless grassroots groups and local park associations that already hold together much of the country's conservation fabric.

Like it or not, the burden of preserving much of Europe's remaining biodiversity now rests on Italy's shoulders. Done well, rewilding and natural capital projects could both safeguard ecosystems and sustain local livelihoods, slowing the rural depopulation that threatens so much of the Apennines. At a time when urban populations across the continent are increasingly starved of contact with the natural world, the wild heart of Italy is an asset of immeasurable value.

I was struggling to get to sleep. The voices in my head were getting louder and I needed some air. I got out of bed and stepped outside. Hooves padded fast and moved off, clipping their way into the distance. A wolf howled, then trailed away. Thankfully none of the rusting streetlights were working. I looked up and saw the stars – millions of them. This in itself is now a rarity in Italy, which ranks among the most light-polluted countries in Europe on a per-capita basis.

Of all the ecological challenges we face, light pollution is one of the simplest to address, and one that offers almost immediate gains: for night-time biodiversity, for the circadian rhythms that govern human health, and for the vast amounts of energy currently wasted illuminating empty roads, office blocks and skies. Italy, after all, possesses a world-renowned lighting-design industry. The problem is not a lack of expertise, but a failure of imagination – and our fear of darkness.

I yawned and returned to bed, finally ready for some sleep.

This beautiful country is both young geologically and young politically, yet ancient in its human history and cultural memory. So old in civilisation, yet so new in its modern form; so rich in natural and cultural wealth, yet so poor in cohesion; so individually brilliant, yet so collectively fragile. It has been shaped by every force that sustains life: by stone, which raised its mountains; by fire, which forged its fertility; by water, which carved its valleys and fed its plains; by islands, which taught the lessons of separation and return; by ice, which tested its endurance; by wood, which sheltered and renewed it; by refuge, which offered safety to both creatures and the preservation of knowledge; and through that, the revival of ideas, revealing the nature of things that led to the Renaissance and the formalisation of natural history. Out of all this upheaval came what defines Italy above all else: abundance – of species, of flavours, of tongues and cultures, of ways to live and imagine the world.

In the end, everything returns to the ground that made this country. Mountains rise and crumble, valleys widen, forests reclaim terraces where ploughs once passed; all of it shaped by the slow, unbroken labour of geology. The Italy we walk upon is not a survivor of the past but a work still in progress, shifting beneath our feet, remaking itself grain by grain. For all our stories and schemes, the land remains the steadier intelligence. It will endure beyond us, as it always has, waiting for whatever comes next.

> *Il bel paese ch'Appennin parte e 'l mar circonda*
> *et l'Alpe.*
> The beautiful country that the Apennines
> divide, and the sea and Alps surround.
>
> **Francesco Petrarca**, 'Sonnet 146',
> *Canzoniere* (*Rerum Vulgarium Fragmenta*)[2]

Acknowledgements

Though not a trained geologist, palaeontologist or natural historian, I have persevered with my research, and this book only makes sense thanks to a huge cast of professionals who generously shared their knowledge, patience and advice. I am particularly indebted to the Italian scientific community, who have continually proven themselves to be brilliant, supportive and kind in response to my enquiries. The greatest pleasure – aside from colouring in my map of Italy – has been meeting that community of experts and enthusiasts along the way. They are often underfunded, underappreciated and disastrously neglected, but they are among the best Italians I have come to know. The book does little justice to their knowledge, but for whatever light it manages to cast, I have them to thank.

For helping me make sense of deep time, I owe special thanks to Cristiano Dal Sasso, Lorenzo Rook, Massimo Delfino, Steve Brusatte, Tim Flannery, Alessandra Fassio, Michelangelo Bisconti, Victoria Herridge, Marco Priori, Giorgio Manzi, Giovanni Destro Bisol, Costantino Buzi and Elvira Visciola, as well as Simone Foresta and Olga Pingue of the Soprintendenza in Benevento for their assistance with Ciro.

For those working more in the present, I am grateful to Papik Genovesi, Telmo Pievani, Luigi Boitani, Franco Andreone, Renato Spicciarelli, Gaetano Aloise, Giovanni

Amori, Sandro Lovari, Dario Capizzi, Francesco Riga, Mauro Mucedda, Totò Inguscio, Simone Cianfanelli, Massimo Lorenzoni, Fabio Stoch, Fabio Cianferoni, Alessandro Montemaggiori, Francesco Petretti, Francesca Ronchi, Anna Loy, Roberto Sindaco, Fabio Attore, Flavio Monti and Andrea Sforzi. Thanks to the Castori Credenti – Mauro Fattor, Luca Lapini, Maurizio Menicucci, Pier Giuseppe Meneguz, Paolo Molinari, Nicola Bressi, Laura De Donato, Emiliano Mori, Beppe Mazza, Gianluca Liva, Renato Pontarini, Davide Righetti, Paola Tepsich and Andrea Mosini – for keeping me up to date. To Bartolomeo 'Aki' Schirone and Cristiano Manni for helping me see the trees; to Alessandro Marchini for helping me see the stars. I had the honour of being advised on volcanoes by Clive Oppenheimer, who introduced me to Boris Behncke of INGV. Paolo Pennisi guided me around Etna, and Clive also introduced me to Brother Guy Consolmagno at the Specola Vaticana, with whom I spent an illuminating morning.

In Rome, I was fortunate to briefly research and write at the Accademia dei Lincei thanks to Alessandro Romanello. Barbara Saracino and Daniela Fugaro kindly organised my visit to the Biblioteca Alessandrina, and Barbara Di Giorgio welcomed me at the Biblioteca Angelica. Edoardo Micati advised on my trip to the hermitages of the Maiella; Pierluigi Astrologo of Majambiente was our magnificent guide there; and Pasquale Mellino and Lorenzo Palumbo hosted us at Terrae Eremis. My thanks also to Daniele Cecca at Castel Porziano; Barbara Franzetti at ISPRA; Bruno Niola and Iole Esposito at Pollino National Park; Saverio Bianco at La Sila; Luca Lombardi at Aspromonte; Alessandro Rossetti at Monti Sibillini and Luciana Zollo at the National Park of Alta Murgia.

Much respect and thanks to Isabella Dalla Ragione for her time and wisdom. Endless affection for Isabella, Carlo Alberto and William Pratesi, and everyone at WWF – this book is written in remembrance of the great Fulco Pratesi, who sadly passed away while I was working on it. For all those involved in conservation in Italy, too many to name here, I owe special thanks to Raffaella Giugni, Massimiliano Falleri, Laura Gentile, Fabio Galluzzo and everyone at Marevivo; Andrea Morello and all at Sea Shepherd; Paolo Fanciulli, Stefano Picchi, Luana Papetti, Mario Cipollone, Umberto Esposito, Gianluca Damiani, Giacomo Radi, Adam Eagle, Francesco Romito, Mia Canestrini, Gaia Agnello, Ambra Messina, Paul Chatterton and George Marshall. Thanks also to Jade Brudenell, Josephine de Chazournes, Rob Cope, Laura Lopresti and the wider Conservation Collective; and to Paul Lister and Jacob Dykes at The European Nature Trust.

I also want to thank Lorenzo Costa, Matteo Mazzola, Dario Fornara and Massimiliano Miatton for showing me real farming; Franco Fubini for believing in real food; and Zach Weiss and Megan Neary for their brilliance in water. To Bo Stehnmeier, Debbie Hinnigan, Ed Maddocks, Alessio Bariviera and Paolo Sodi – it was a wild ride regardless. My thanks as well to Luigi De Vecchi, Nino Tronchetti Provera, Riccardo and Elizabeth Orizio, Alessandra Galloni, Andrea Scarpa, Salvo Barrano, Sammy Leslie, Jason Horowitz and Chris Morgan for early advice and support.

Thank you above all to my editor, Muna Reyal, for taking a chance on a first-time author and believing in this project from the start. To Susan Pegg for her eyes of a hawk, and priceless reassurance. To the whole Chelsea Green team, my gratitude. Thanks also to Stephen Spurr for being a constant ear and inspiration.

To all my loyal friends – too many to name in full – but especially Ben Goldsmith, Euan Davidson, Iain Russell, James Sevier, Ben Whately, Giles Hayward, Josh Horsfield, Tom Hamilton, William Aitken, James Hooper, George Morton-Jack, Alex Tulloch, Ed and Sophia Sackville, Tristan Powell, Ade, Marc De Puechredon, Leo and Claire de Montaignac, Leo Dugdale, Will Palau, Simona, Nicole and Maria Vittoria Grimaudo, Maria Elena Barnabi, James Graham, Johnny and Liz Petrucci, Luca and Giorgia Savelli, Alberto and Francesca Tasca and Cosmin User for friendship, fixing my lens, and much else over the years.

Finally, to my family, without whom none of this would have been possible: John & Kate, Kate & Stephen; to Elvira, Carla, Fulvio Kripal and Angelo. Above all my wife, Isabella, and our three amazing children, Franco, Costanza and Victoria, who have endured more than the usual amount of chaos these past years; and ultimately my parents, Tim and Christine, who have never stopped supporting me in whatever I do. This is for them.

Bibliography and Further Reading

Abulafia, David. *The Great Sea: A Human History of the Mediterranean*. Allen Lane, 2011.

Alvarez, Walter. *T. rex and the Crater of Doom*. Princeton University Press, 1997.

Alvarez, Walter. *The Mountains of Saint Francis: Discovering the Geologic Events That Shaped Our Earth*. W.W. Norton & Company, 2008.

Attlee, Helena. *The Land Where Lemons Grow: The Story of Italy and Its Citrus Fruit*. Penguin Books, 2014.

Barron, Patrick, and Anna Re, eds. *Italian Environmental Literature: An Anthology*. Italica Press, 2003.

Barzini, Luigi. *The Italians: A Full-Length Portrait Featuring Their Manners and Morals*. Atheneum, 1964.

Benton, Michael J. *Extinctions: How Life Survives, Adapts and Evolves*. Thames & Hudson, 2023.

Bjornerud, Marcia. *Timefulness: How Thinking Like a Geologist Can Help Save the World.* Princeton University Press, 2018.

Boitani, Luigi. *Dalla parte del lupo*. Mondadori, 1992.

Boitani, Luigi, and Roger A. Powell, eds. *Carnivore Ecology and Conservation: A Handbook of Techniques*. Oxford University Press, 2012.

Boitani, Luigi, and Paul C. Paquet, eds. *Large Carnivores and the Conservation of Biodiversity*. Island Press, 2004.

Bonatti, Walter. *The Mountains of My Life*. Translated by Robert Marshall. Arrow Books, 2010.

Bosellini, Alfonso. *Storia geologica d'Italia: Gli ultimi 200 milioni di anni.* Zanichelli, 2005.

Brannen, Peter. *The Ends of the World: Volcanic Apocalypses, Lethal Oceans, and Our Quest to Understand Earth's Past Mass Extinctions.* Ecco, 2017.

Brusatte, Steve. *The Rise and Fall of the Dinosaurs: A New History of a Lost World.* William Morrow, 2018.

Brusatte, Steve. *The Rise and Reign of the Mammals: A New History, from the Shadow of the Dinosaurs to Us.* Mariner Books, 2022.

Burckhardt, Jacob. *The Civilization of the Renaissance in Italy.* Translated by S.G.C. Middlemore. Penguin Classics, 1990.

Calvino, Italo. *The Baron in the Trees.* Translated by Archibald Colquhoun. Random House, 1959.

Calvino, Italo. *Marcovaldo: Or The Seasons in the City.* Translated by William Weaver. Vintage Classics, 2023.

Carrada, Giovanni, ed. *L'altro Rinascimento. Ulisse Aldrovandi e le meraviglie del mondo.* Bologna University Press, 2022.

Corona, Mauro. *Le voci del bosco.* Mondadori, 1998.

Cotugno, Ferdinando. *Italian Wood.* Mondadori Electa, 2020.

Cristofani, Mauro. *The Etruscans.* Rizzoli, 1992.

Dalla Ragione, Livio, and Isabella Dalla Ragione. *Arboreal Archeology: A Diary of Two Fruit Explorers.* Translated by Cecilia Galiena. Ali & No Editrice, 2008.

Dalla Ragione, Isabella. *Frutti ritrovati: 100 varietà antiche e rare da scoprire.* Mondadori, 2010.

Dal Prete, Ivano. *On the Edge of Eternity: The Antiquity of the Earth in Medieval and Early Modern Europe.* Oxford University Press, 2022.

Dal Sasso, Cristiano. *I dinosauri spiegati a mio figlio.* Rizzoli, 2007.

Destro Bisol, Giovanni, and Marco Capocasa. *Italiani: Come il DNA ci aiuta a capire chi siamo.* Vol. 13 of *Città della scienza.* Carocci, 2016.

Dickie, John. *Delizia! The Epic History of the Italians and Their Food.* Sceptre, 2007.

Dreher, Rod. *The Benedict Option: A Strategy for Christians in a Post-Christian Nation*. Sentinel, 2017.

Eco, Umberto. *The Name of the Rose*. Translated by William Weaver. Harcourt Brace Jovanovich, 1983.

Eco, Umberto. *On Beauty: A History of a Western Idea*. Secker & Warburg, 2004.

Flannery, Tim. *Europe: The First 100 Million Years*. Penguin, 2019.

Freedberg, David. *The Eye of the Lynx: Galileo, His Friends, and the Beginnings of Modern Natural History*. University of Chicago Press, 2002.

Gee, Henry. *A (Very) Short History of Life on Earth: 4.6 Billion Years in 12 Pithy Chapters*. St. Martin's Press, 2021.

Gee, Henry. *The Decline and Fall of the Human Empire: A Story of Our Future, Autumn AD 10,001*. St. Martin's Press, 2025.

Gilmour, David. *The Pursuit of Italy: A History of a Land, Its Regions, and Their Peoples*. Allen Lane, 2011.

Gordon, Helen. *Notes from Deep Time: A Journey Through Our Past and Future Worlds*. Profile Books, 2021.

Gould, Stephen Jay. *Wonderful Life: The Burgess Shale and the Nature of History*. W.W. Norton, 1989.

Greenblatt, Stephen. *The Swerve: How the World Became Modern*. W.W. Norton & Company, 2011.

Gribbin, John, and Mary Gribbin. *On the Origin of Evolution: Tracing 'Darwin's Dangerous Idea' from Aristotle to DNA*. William Collins, 2020.

Halliday, Thomas. *Otherlands: A Journey Through Earth's Extinct Worlds*. Allen Lane, 2022.

Harper, Kyle. *The Fate of Rome: Climate, Disease, and the End of an Empire*. Princeton University Press, 2017.

Hooper, John. *The Italians*. Viking, 2015.

Jacobsen, Rowan. *Truffle Hound: On the Trail of the World's Most Seductive Scent, with Dreamers, Schemers, and Some Extraordinary Dogs*. Bloomsbury Publishing, 2021.

Jones, Tobias. *The Dark Heart of Italy: Travels Through Time and Space Across Italy*. Faber & Faber, 2003.

Jones, Tobias. *Utopian Dreams: A Search for a Better Life*. Faber & Faber, 2007.

Jones, Tobias. *The Po: An Elegy for Italy's Longest River*. Apollo, 2022.

King, Ross. *Michelangelo and the Pope's Ceiling*. Pimlico, 2003.

Lawrence, D.H. *D.H. Lawrence and Italy: Sketches from Etruscan Places, Sea and Sardinia, Twilight in Italy*. Edited by Simonetta de Filippis et al. Penguin Classics, 2008.

da Vinci, Leonardo. *Notebooks*. Selected by Irma A. Richter, edited with an introduction and notes by Thereza Wells. Oxford University Press, 2008.

Levi, Carlo. *Christ Stopped at Eboli*. Translated by Frances Frenaye. Penguin Classics, 2000.

Long, John A. *The Secret History of Sharks: The Rise of the Ocean's Most Fearsome Predators*. Ballantine Books, 2024

Lucretius. *The Nature of Things*. Translated by A.E. Stallings. Penguin Classics, 2007.

Macfarlane, Robert. *Underland: A Deep Time Journey*. W.W. Norton & Company, 2019.

Macfarlane, Robert. *Mountains of the Mind: A History of a Fascination*. Granta Books, 2003.

Macfarlane, Robert. *Is a River Alive?* W.W. Norton & Company, 2025.

Mancuso, Stefano. *Brilliant Green: The Surprising History and Science of Plant Intelligence*. Translated by Joan Benham. Island Press, 2015.

Mancuso, Stefano. *The Revolutionary Genius of Plants: A New Understanding of Plant Intelligence and Behavior*. Atria Books, 2018.

Mancuso, Stefano. *The Nation of Plants*. Translated by Gregory Conti. Other Press, 2021.

Manzi, Giorgio. *L'ultimo Neanderthal racconta: Storie prima della storia*. Mondadori, 2009.

Manzi, Giorgio. *Antenati: Lucy e altri racconti dal tempo profondo.* Mondadori, 2012.

Micati, Edoardo. *Eremi d'Abruzzo. Guida ai luoghi di culto rupestri.* CARSA Edizioni, 2021.

Michieli, Franco. *La vocazione di perdersi: Piccolo saggio su come le vie trovano i viandanti.* Ediciclo Editore, 2015.

Militello, Cristiano. *Giulietta è 'na zoccola: Gli striscioni più esilaranti delle curve italiane.* Kowalski, 2004.

Oppenheimer, Clive. *Eruptions That Shook the World.* Cambridge University Press, 2011.

Oppenheimer, Clive. *Mountains of Fire: The Secret Lives of Volcanoes.* Hodder & Stoughton, 2023.

Pedrotti, Franco. *I pionieri della protezione della natura in Italia.* Tipolitografia Editrice TEMO, 2012.

Petrini, Carlo. *Slow Food Nation: Why Our Food Should Be Good, Clean, and Fair.* Translated by Clara Furlan and Jonathan Hunt. Rizzoli Ex Libris, 2013.

Petrini, Carlo. T*erra Madre: Forging a New Global Network of Sustainable Food Communities.* Translated by Jonathan Hunt. Chelsea Green Publishing, 2009.

Pievani, Telmo, and Mauro Varotto. *Viaggio nell'Italia dell'Antropocene: La geografia visionaria del nostro futuro.* Aboca Edizioni, 2021.

Pievani, Telmo. *La vita inaspettata: Il fascino di un'evoluzione che non ci aveva previsto.* Raffaello Cortina Editore, 2011.

Pievani, Telmo. *Imperfezione: Una storia naturale.* Raffaello Cortina Editore, 2019.

Pliny the Elder. *Natural History: A Selection.* Translated by John F. Healy. Penguin Books, 1991.

Pratesi, Fulco. *Storia della natura d'Italia.* Editori Riuniti, 2001.

Rigoni Stern, Mario. *Uomini, boschi e api.* Edited with an introduction by Erri De Luca. Einaudi, 2021.

Rosenblum, Mort. *Olives: The Life and Lore of a Noble Fruit.* North Point Press, 1996.

Rumiz, Paolo. *La leggenda dei monti naviganti*. Feltrinelli, 2007.
Shelley, Percy Bysshe. 'Julian and Maddalo: A Conversation.' In *The Complete Poetical Works of Percy Bysshe Shelley*, edited by Thomas Hutchinson. Clarendon Press, 1904.
Spencer, Nicholas. *Magisteria: The Entangled Histories of Science and Religion*. Oneworld Publications, 2024.
Spicciarelli, Renato. *La Psiche del Frassino*. Consiglio Regionale della Basilicata – Finiguerra Arti Grafiche, 2004.
Stoppani, Antonio. *Il bel paese: Conversazioni sulle bellezze naturali, la geologia e la geografia fisica d'Italia*. Giacomo Agnelli, 1876.
Stuart, Anthony J. *Vanished Giants: The Lost World of the Ice Age*. University of Chicago Press, 2021.
Svevo, Italo. *Zeno's Conscience*. Translated by William Weaver. Vintage, 2003.
Sweeney, Jon M. *The Pope Who Quit: A True Medieval Tale of Mystery, Death, and Salvation*. Image Books, 2012.
Viestad, Andreas. *Dinner in Rome: A History of the World in One Meal*. Reaktion Books, 2022.
Virgil. *The Georgics*. Translated by L.P. Wilkinson. Penguin, 1982.
Zovi, Daniele. *Italia selvatica: Storie di orsi, lupi, gatti selvatici, cinghiali, lontre, sciacalli dorati, linci e un castoro*. UTET, 2019.

Notes

Introduction: *Salamandrina*

1. I also began to understand the differentiation made between a species and subspecies, often the focus of continuing debate but ultimately defined by close genetic study performed within the scientific community. For the sake of simplicity, I don't enter such debate here but include both categories for their endemic status, trying where possible to make the distinction.
2. Tim Flannery, *Europe: The First 100 Million Years* (Penguin, 2019), 34.
3. *Il Bel Paese* sold six hundred thousand copies upon publication in 1876, probably about equal to the number of people who were literate in Tuscan (modern) Italian. At the time in which the book was written, the vast majority of the population still spoke one of about thirty different dialects and languages present across Italy. It went on to become a compulsory text in all Italian schools and a defining work for the newly unified country, although it's less used today. Fulco Pratesi was the beloved founder of the World Wide Fund for Nature (WWF) in Italy and the godfather of Italian conservation. I had the pleasure and pride of knowing him, and I felt an immense loss when he passed in the midst of writing this book.

Chapter One: Stone

1. Giovanni Todesco, *Incredibile! Ho Trovato il Primo Dinosauro Italiano* (unpublished manuscript, Pietraroja, 2001), 2, https://www.dinosauroantonio.it/doc/incredibile_ho_trovato_il_primo_dinosauro_italiano.pdf.
2. Todesco, *Incredibile!*, 3.
3. Cristiano Dal Sasso and Simone Maganuco, '*Scipionyx samniticus* (Theropoda: Compsognathidae) from the Lower Cretaceous of Italy: Osteology, Ontogenetic Assessment, Phylogeny, Soft Tissue Anatomy,

Taphonomy, and Palaeobiology', *Memorie della Società Italiana di Scienze Naturali e del Museo Civico di Storia Naturale di Milano* 37, no. 1 (2011): 1–281, https://www.researchgate.net/publication/231347243_Scipionyx_samniticus_Theropoda_Compsognathidae_from_the_Lower_Cretaceous_of_Italy_Osteology_ontogenetic_assessment_phylogeny_soft_tissue_anatomy_taphonomy_and_paleobiology.

4. *Scipionyx* refers to both the geologist Scipione Breislak (1748–1826), who first described the Pietraroja site, and Scipio Africanus, the renowned Roman general who defeated Hannibal. Onyx is from the Greek for claw, and *samniticus* is Latin for 'of the Samnium', the ancient upland region of southern Italy inhabited by the Samnites. The name *Scipionyx samniticus* thus means 'Scipio's claw of the Samnium'.
5. The Adria microplate (or Adriatic plate) is a small fragment of the African plate that broke away tens of millions of years ago. As it drifted northward and collided with Europe, its leading edge was forced beneath the continent, crumpling and uplifting the overlying sediments to form the Apennines and parts of the Alps.
6. Walter Alvarez and David H. Shimabukuro, 'The Geological Relationships Between Sardinia and Calabria During Alpine and Hercynian Times', *Italian Journal of Geosciences* 128, no. 2 (2009): 257–268, https://doi.org/10.3301/IJG.2009.128.2.257.

Chapter Two: Fire

1. Francesco G. Fedele, Biagio Giaccio, Roberto Isaia and Giovanni Orsi, 'The Campanian Ignimbrite Eruption, Heinrich Event 4, and Palaeolithic Change in Europe: A High-Resolution Investigation', in *Volcanism and the Earth's Atmosphere*, eds A. Robock and C. Oppenheimer (American Geophysical Union, 2003), 301–325, https://doi.org/10.1029/139GM20.
2. Cappellaccio tuff is among the oldest gifts from the Alban eruptions, dating to five hundred thousand years ago. It was quarried along the Tiber valley, at the base of the Palatine and the southern edge of Rome in its earliest foundation years, and is found in all the lowest layers of the city.
3. Dunn, Rob, 'Has a New 10-Legged Species Evolved beneath Rome?', *Scientific American* (guest blog), 14 June 2012, https://www.scientificamerican.com/blog/guest-blog/has-a-new-ten-legged-species-evolved-beneath-rome.

4. Humberto Basilio, 'Pompeii Time Capsule Reveals Secrets to Durable Ancient Roman Cement', *Scientific American*, 9 December 2025, https://www.scientificamerican.com/article/pompeii-house-frozen-mid-renovation-reveals-secrets-of-roman-cement.
5. Linda M. Seymour et al., 'Hot Mixing: Mechanistic Insights into the Durability of Ancient Roman Concrete', *Scientific Advances* 9, no. 1 (2023): eadd1602, https://doi.org/10.1126/sciadv.add1602.
6. Basalt is the result of effusive eruptions, when low-viscosity magma flows steadily across the surface rather than exploding into ash and pumice. Though the Alban Hills were best known for their violent, explosive phases, they also produced occasional basaltic lava flows, now quarried for paving and building stone throughout the Roman countryside.
7. Renato Spicciarelli, 'Update on Aspects of Ethoecology for the Exceedingly Restricted and Most Outstanding Italian Endemic Moth, *Brahmaea (Acanthobrahmaea) europaea* Hartig (Lepidoptera: Brahmaeidae)', *Fragmenta Entomologica* 54, no. 2 (2022): 311–314, https://doi.org/10.13133/2284-4880/1387; Fabio Mosconi et al., 'An Overview on the Most Outstanding Italian Endemic Moth, *Brahmaea (Acanthobrahmaea) europaea* (Lepidoptera: Brahmaeidae)', *Fragmenta Entomologica* 46, no. 1–2 (2014): 1-9. https://doi.org/10.4081/fe.2014.70.
8. Paolo Mietto et al., eds., '2001–2021: Vent'anni di Ricerche sulle "Ciampate del Diavolo". Dalla Leggenda alla Realtà Scientifica', *Miscelllania Istituto Nazionale di Geofisica e Vulcanologia* 64 (2022): 1–248, https://doi.org/10.13127/misc/64.
9. Marisa Giuffrida et al., 'Etna 2011–2022: Discoveries from a Decade of Activity at the Volcano', *Earth-Science Reviews* 245 (2023): 104563, https://doi.org/10.1016/j.earscirev.2023.104563.

Chapter Three: Water

1. Lorenzo Lipparini et al., 'Extensive Freshened Groundwater Resources Emplaced During the Messinian Sea-level Drawdown in Southern Sicily, Italy', *Communications Earth & Environment* 4 (2023): 430, https://doi.org/10.1038/s43247-023-01077-w.
2. Marco Romano and Giorgio Carnevale, 'The Early Studies on the Eocene Bolca Fossil-Lagerstätte (Italy): An Historical Overview', *Bollettino della Società Paleontologica Italiana* 62, no. 2 (2023): 105–142, https://doi.org/10.4435/BSPI.2023.05; Matt Friedman

and Giorgio Carnevale; 'The Bolca Lagerstätten: Shallow Marine Life in the Eocene', *Journal of the Geological Society* 175, no. 4 (2018): 569–579, https://doi.org/10.1144/jgs2017-164.

3. Giovanni Bianucci et al., 'Rise of the Titans: Baleen Whales Became Giants Earlier than Thought', *Biology Letters* 15, no. 5 (2019): 20190175, https://doi.org/10.1098/rsbl.2019.0175;.Michelangelo Bisconti et al., 'Il Patrimonio Paleocetologico Piemontese: Storia E Prospettive', *Parva Naturalia* 15 (2020): 15–58, https://www.museogeologico.it/images/stories/parva/parva_naturalia_2020.pdf; Sara Citron et al., 'Systematics, Phylogeny and Feeding Behavior of the Oldest Killer Whale: A Reappraisal of *Orcinus citoniensis* (Capellini, 1883) from the Pliocene of Tuscany (Italy)', *Bollettino della Società Paleontologica Italiana* 61, no. 2 (2022): 167–186, https://www.paleoitalia.it/wp-content/uploads/2022/09/06_Citron_et_al_2022_BSPI_612.pdf.
4. Kay Etheridge, 'Leonardo and the Whale', in *Leonardo da Vinci – Nature and Architecture*, ed. Constance Moffat and Sara Taglialagamba (Brill, 2019), 89–106.
5. Etheridge, 'Leonardo and the Whale', 89–106.
6. Isobell Cockerell, 'The Secret Movement Bringing Europe's Wildlife Back from the Brink', Coda, 27 June 2023, https://www.codastory.com/climate-crisis/rewilding-beavers-conservation.
7. 'Siti di interesse nazionale (SIN), Istituto Superiore per la Protezione e la Ricerca Ambientale (ISPRA), last modified June 2025, https://www.isprambiente.gov.it/it/attivita/suolo-e-territorio/siti-contaminati/siti-di-interesse-nazionale-sin.
8. Dressman, E. Chase, 'Italian Court Sentences 11 Over Crimes Related to PFAS Contamination', Law Bulletins, Taft Law, 3 July 2025, https://www.taftlaw.com/news-events/law-bulletins/italian-court-sentences-11-over-crimes-related-to-pfas-contamination.

Chapter Four: Islands

1. Seven *Podarcis* species are native to Italy: the common wall lizard (*P. muralis*), Italian wall lizard (*P. siculus*), Tyrrhenian wall lizard (*P. tiliguerta*), Sicilian wall lizard (*P. waglerianus*), Aeolian wall lizard (*P. raffonei*), Maltese wall lizard (*P. filfolensis*, on Lampedusa and Linosa) and the Dalmatian wall lizard (*P. melisellensis*, on a few Adriatic islets). Altogether, these comprise more than forty

described subspecies, many restricted to single islands or localities. (Roberto Sindaco and Edoardo Razzetti, 'An Updated Check-List of Italian Amphibians and Reptiles', *Natural History Sciences* 8, no. 2 (2021): 35–46, https://doi.org/10.4081/nhs.2021.519.)

2. Patrick Barkham, 'Wall Lizards Reveal How Evolution Can Happen in Real Time', Guardian, 28 February 2026, https://www.theguardian.com/environment/2026/feb/28/conservation-biodiversity-species-genetics-wall-lizards-aeolian-italy-aoe.
3. Matthijs Freudenthal et al., 'When and How Did the *Mikrotia* Fauna Reach Gargano (Apulia, Italy)?' *Geobios* 46, no. 1–2 (2013): 105–109, https://doi.org/10.1016/j.geobios.2012.10.004.
4. It is believed the unusual fauna survived on Gargano between eleven and four million years ago, enduring even the Messinian Salinity Crisis and the refilling of the Mediterranean. Afterward, the island appears to have been submerged, before being uplifted again a million years later and bolted onto the rest of Italy as it rose above the waves. The most vivid modern recreation is by Thomas Halliday in his brilliant book *Otherlands* (Random House, 2022).
5. *Oreopithecus bambolii* is on permanent display at the Museo di Storia Naturale of the University of Florence. A detailed reconstruction of Sandrone can also be seen at the Museo di Storia Naturale della Maremma in Grosseto.
6. The anthropoid apes (infraorder Simiiformes) spread widely across Eurasia during the Miocene. While *Oreopithecus* itself left no descendants, its wider group – the hominoids – survives today in the great apes, including orangutans, gorillas, chimpanzees, bonobos and humans.
7. Daniel DeMiguel and Lorenzo Rook, 'An Updated Overview on the Extinction of *Oreopithecus bambolii*', *Bollettino Della Società Paleontologica Italiana* 63 (2024): 183–192, https://hdl.handle.net/2158/1374532.
8. Sardinia hosts five of Europe's eight cave salamander species, all belonging to the genus *Speleomantes* (formerly *Hydromantes*): Gené's cave salamander (*Speleomantes genei*), Monte Albo cave salamander (*Speleomantes flavus*), imperial cave salamander (*Speleomantes imperialis*), Supramonte cave salamander (*Speleomantes supramontis*), and Sarrabus or Sette Fratelli cave salamander (*Speleomantes sarrabusensis*).
9. All seven of the Aeolian islands, Lipari, Alicudi, Filicudi, Salina, Panarea, Vulcano and Stromboli, are volcanic and although most are quiescent, with only Lipari, Stromboli and Vulcano considered active, all have the

potential to reawaken. Filicudi is the oldest, going back a million years, but the rest emerged at various times between 10,000 and 400,000 years ago. Less known are the six underwater volcanoes that are part of the same chain – Sisifo, Enarete, Eolo, Lametini, Alcione and Palinuro.

Chapter Five: Ice

1. Qiaomei Fu et al., 'The Genetic History of Ice Age Europe', *Nature* 534, no. 7606 (2016): 200–205. https://doi.org/10.1038/nature17993.
2. Johannes Reinthaler and Frank Paul, 'Reconstructed Glacier Area and Volume Changes in the European Alps Since the Little Ice Age', *The Cryosphere* 19, no. 2 (2025): 753–67, https://doi.org/10.5194/tc-19-753-2025.

Chapter Six: Wood

1. CIL X.6225 (Theodore Mommsen, ed., *Corpus Inscriptionum Latinarum X: Inscriptiones Bruttiorum, Lucaniae, Campaniae, Siciliae, Sardiniae Latinae* (apud G. Reimerum, 1883), 6225.). Inscription from Minturno (AD 249) commemorating a festival in which 40 bears and 16 deer were killed.
2. Pliny the Elder, *Natural History*, trans. H. Rackham, Loeb Classical Library 371 (Harvard University Press, 1949), 18.7. Latin: '*rapiuntur magis quam coluntur*' ('they are plundered rather than cultivated').
3. Antonello Venditti, 'Notte prima degli esami', on *Cuore*, Heinz Music,1984.
4. Giovanni de Giudici et al., 'Application of Hydrologic-Tracer Techniques to the Casargiu Adit and Rio Irvi (SW-Sardinia, Italy): Using Enhanced Natural Attenuation to Reduce Extreme Metal Loads', *Applied Geochemistry* 96 (2018): 42–54, https://doi.org/10.1016/j.apgeochem.2018.06.004.

Chapter Seven: Refuge

1. Hugh Ford, 'St. Benedict of Norcia', *The Catholic Encyclopedia*, vol. 2 (Robert Appleton Company, 1907).
2. Pliny the Elder's more absurd claims: Geese from driftwood or seashells – discussed in *Book XI*, his treatment of land animals, where he records spontaneous generation theories; bees springing from the ox's carcass – mentioned in *Book XI*, within his discourse on insects and their mysterious origins; proclaiming ostriches can

digest metal – featured in *Book VIII*, his section on land animals and exotic creatures; dog-headed men who barked – part of *Book VII* on anthropology and human physiognomy, where mythical tribes are detailed; the basilisk myth – Pliny's description appears in *Book VIII*, about fantastic serpents, presenting it as the first Western record of the basilisk myth.

3. Jon M. Sweeney, *The Pope Who Quit: A True Medieval Tale of Mystery, Death, and Salvation* (Image Books, 2012).
4. Francesco Petrarca, *Letters on Familiar Matters* (*Rerum Familiarium Libri* 3.9), trans. Aldo S. Bernardo (Albany: State University of New York Press, 1975), 158–160.
5. 'La Flora del Parco', Parco Nazionale della Maiella, last modified 24 June 2020, https://www.parcomajella.it/La-Flora-del-Parco.htm.
6. Guida alle orchidee del Parco Nazionale della Majella, ed. M. Biagioli, P. L. Nimis, M. Di Cecco, A. Moro, S. Martellos, E. Pittao, and G. Ciaschetti (Trieste: Università degli Studi di Trieste, Progetto Dryades, 2023), available online - http://dryades.units.it/online_books/Orchidee_Majella.pdf?
7. Marco D'Antraccoli, Roberta Vangelisti, Roberta and Lorenzo Peruzzi, 'Luca Ghini and the Origin of Modern Botany: An Italian History of Academic Botanic Gardens', in *2nd International Congress of Historical Botanical Gardens*, 29–31 July 2024, Vienna, https://www.historicalbotanicalgardens.com/wp-content/uploads/abstracts.pdf.
8. Gian Battista Vai and William Cavazza, 'Ulisse Aldrovandi and the Origin of Geology and Science', in *The Origins of Geology in Italy*, ed. Gian Battista Vai, W. Glen and E. Caldwell (Geological Society of America, 2006), 43–63, https://doi.org/10.1130/2006.2411(04).
9. Mattioli's work was published under the full title *I discorsi ne i sei libri della materia medicinale di Pedacio Dioscoride Anazarbeo* (first published 1544).
10. Andrew Wear, 'Medicine in Early Modern Europe', in *The Western Medical Tradition, 800–1800*, ed. Lawrence I. Conrad et al. (Cambridge University Press, 1995), 304.
11. In the Biblioteca Angelica in Rome, Gherardo Cibo's Herbarium is preserved – the oldest of its kind in the world. Beginning in the 1530s, Cibo pressed and glued plants onto paper using a technique taught to him by Luca Ghini. The pages remain astonishing today: plants dried nearly five centuries ago are still instantly recognisable, many retaining traces of their original colour. They resemble the

finest fossils from Bolca or Pietraroja, perfectly 'flash-preserved', as if freshly cut only days ago.

Chapter Eight: Revival

1. A. Sfriso et al., 'Spreading of Alien Macroalgae in the Venice Lagoon, the Italian Hotspot of Non-Indigenous Species: Biodiversity and Standing Crop', *Biologia Marina Mediterranea* 22, no. 1 (2015): 163–166.
2. Spike Bucklow, *The Alchemy of Paint: Art, Science and Secrets from the Middle Ages* (Marion Boyars, 2009); Victoria Finlay, *Color: A Natural History of the Palette* (Random House, 2002); Kelly Grovier, *The Art of Colour: The History of Art in 39 Pigments* (Thames & Hudson, 2023).
3. Ross King, *Michelangelo and the Pope's Ceiling: The Undiscovered Story of the Sistine Chapel* (Walker & Co., 2003).

Chapter Nine: Abundance

1. Ivan Serafini, 'Asparagi, le 8 Specie Presenti in Italia: Come Riconoscerli, Raccoglierli e Coltivarli', Centro Studi Alto Vastese e Valle del Trigno, 26 March 2013, https://www.altovastese.it/flora-2/asparagi-le-8-specie-presenti-in-italia-come-riconoscerli-raccoglierli-e-coltivarli.
2. Pliny the Elder, *Natural History*, Book XX, Section 86 (*De brassica*). See: *Naturalis Historia*, XX.86: '*urinam eorum, qui brassicam diu ederint, tepide inlinendam in omnium membrorum dolorem... pueri lavati ea numquam debilitantur.*'
3. Francesca Oliva, 'Svelare il Mistero e la Magia che Circonda l'Olivo della Strega in Toscana', Olive Oil Times, 4 June 2023, https://it.oliveoiltimes.com/business/europe/unveiling-the-mystery-and-magic-surrounding-tuscanys-olivo-della-strega/120422.
4. Chiara Sabelli, 'Deadly Olive Tree Pathogen Came by Road and Rail', *Nature*, 17 August 2023, https://doi.org/10.1038/d43978-023-00118-4.
5. Dan Saladino, 'Bitterness', *The Food Programme*, BBC Radio 4, 4 October 2015, https://www.bbc.co.uk/food/programmes/b06f4z3k.
6. Lanfredo Castelletti, 'La Domesticazione delle Piante e l'Agricoltura in Europa Preistorica e Protostorica', *Il Mondo dell'Archeologia* (Istituto della Enciclopedia Italiana fondata da Giovanni Treccani, 2002), https://www.treccani.it/enciclopedia/la-domesticazione-delle-piante-e-l-agricoltura-europa-preistorica-e-protostorica_(Il-Mondo-dell%27Archeologia).

7. Minor M. Markle III, 'The Macedonian Sarrissa, Spear and Related Armor', *American Journal of Archaeology* 81, no. 3 (1977): 323–39, https://doi.org/10.2307/503007.
8. Species referred to include: the edible chanterelle (*Cantharellus cibarius*), porcini (*Boletus edulis*), Caesar's mushroom (*Amanita caesarea*) and horn of plenty (*Craterellus cornucopioides*); and the highly toxic and inedible death cap (*Amanita phalloides*), panther cap (*Amanita pantherina*), destroying angel (*Amanita virosa*) and deadly webcaps (*Cortinarius* spp.).
9. FAO, *State of the World's Biodiversity for Food and Agriculture*, ed. J. Bélanger & D. Pilling, (FAO Commission on Genetic Resources for Food and Agriculture Assessments, 2019); ISPRA, La biodiversità agricola in Italia; Slow Food Foundation for Biodiversity. Italy is repeatedly identified as a hotspot for agrobiodiversity, especially for the density of cultivated varieties and landraces per unit area.
10. Giorgio Ghiglione, 'Underwater Museum: How "Paolo the Fisherman" Made the Med's Strangest Sight', *Guardian*, 17 November 2020, https://www.theguardian.com/environment/2020/nov/17/underwater-museum-how-paolo-the-fisherman-made-the-meds-strangest-sight.
11. Paolo Fanciulli, 'Pescaturismo a Talamone', https://paoloilpescatore.it.
12. Italy is unique in that farro cultivation, particularly of emmer wheat (*Triticum dicoccum*), is not only well-established but also expanding, in contrast to its decline elsewhere. In the mountainous Garfagnana region of Tuscany, emmer is grown under an IGP (Indicazione Geografica Protetta) designation, granting it legal protection based on geographic origin. The rising demand for Italian farro has led to competition from uncertified or industrially produced varieties, often grown in lowland areas and based on different wheat species such as spelt (*Triticum spelta*).
13. Starnini, Elisabetta, Paolo Biagi and Niccolò Mazzucco, 'The Beginning of the Neolithic in the Po Plain (Northern Italy): Problems and Perspectives', *Quaternary International* 470, Part B (2018): 301–17, https://doi.org/10.1016/j.quaint.2017.05.059.
14. Elvira Visciola, ed., 'TwoHeaded Statue of Campo Ceresole – Vhò di Piadena (CR)', Prehistory in Italy, accessed 3 September 2025, https://www.preistoriainitalia.it/en/scheda/statuina-bicefala-di-campo-ceresole-vho-di-piadena-cr.

15. Mark B. Tauger, 'Nazareno Strampelli and the First Green Revolution', *Journal of International Development* 35, no. 8 (2023): 2485–2512, https://doi.org/10.1002/jid.3785.; Michele Sollai, 'The Fascist Green Revolution', *Plants, People, Planet* 6, no. 5 (2024): 1094–110, https://doi.org/10.1002/ppp3.10386; Martina Ludovica Careddu et al., 'Lessons from the Varietal Evolution of Durum Wheat in Italy', *Agronomy* 14, no. 1 (2024): 87, https://doi.org/10.3390/agronomy14010087; S. Salvi et al., 'Nazareno Strampelli, the "Prophet" of the Green Revolution', *The Journal of Agricultural Science* 151, no. 1 (2013): 1–5, https://doi.org/10.1017/S0021859612000214.

Epilogue: The Nature of Things

1. 'Aree Protette: Decreto Pichetto, Nasce il 25mo Parco Nazionale, è Quello del Matese', Ministero dell'Ambiente e della Sicurezza Energetica, 22 April 2025, https://www.mase.gov.it/portale/-/aree-protette-decreto-pichetto-nasce-il-25mo-parco-e-quello-del-matese.
2. Francesco Petrarca, *Canzoniere: Selected Poems*, trans. Anthony Mortimer (Penguin Classics, 2002), Sonnet 146, lines 13–14.

Index

Page numbers in **bold** refer to the plates section.

Index

Index

Index

About the Author

Johnny Petrucci

Edward Cutler is a British-Italian writer, organic farmer and environmental advocate based between Rome and southern Tuscany. Born in Britain and educated at the University of Edinburgh, he has lived in Italy since 2009, where his fascination with the country's landscapes, geology and biodiversity gradually drew him away from the art world and into green technology, farming and conservation.

In 2020, together with friends and conservationists, Ed founded the Tuscany Environment Foundation, part of the Conservation Collective, which supports environmental projects across the region.

Today, Ed produces organic olive oil in the Tuscan countryside and works on environmental and rewilding initiatives across Italy. He is also course director of the executive programme Biodiversity and Business Sustainability at the European Institute of Innovation for Sustainability in Rome.

This Beautiful Country is his first book.